Extreme Expeditions

EXTREME EXPEDITIONS

Travel Adventures Stalking the World's Mystery Animals

By Adam Davies

Anomalist Books
*San Antonio * New York*

An Original Publication of ANOMALIST BOOKS

EXTREME EXPEDITIONS
Travel Adventures Stalking the World's Mystery Animals
Copyright © 2008 by Adam Davies
ISBN: 1933665319

Portions of this book have been previous published, in slightly different form, in: "Death Worm!" *Fortean Times*, April 2004; and "I thought I saw a 'Sauropod,'" *Fortean Times*, April 2001.

Front and back cover photos by Andrew Sanderson

Book design by Seale Studios

All rights reserved, including the right to reproduce this book or portions thereof in any form whatsoever.

For information, go to anomalistbooks.com, or write to:
Anomalist Books, 5150 Broadway #108, San Antonio, TX 78209

Contents

Author's Note viii

Chapter 1: The Dream 1

Chapter 2: War in The Congo 6

Chapter 3: World News 13

Chapter 4: The Seljord Serpent 20

Chapter 5: Congo or Bust 37

Chapter 6: The Loch Ness Monster 42

Chapter 7: The Last Great Dinosaur Hunt 51

Chapter 8: Orang-Pendek 90

Chapter 9: Back to Sumatra 119

Chapter 10: The Mongolian Deathworm 124

Chapter 11: Russian Bigfoot 136

Expedition Gallery 149

Acknowledgments 154

Chronology of Expeditions 155

To Alexandria:
Your short life inspired me to live mine to its fullest.
It's because of you I do these things.

Author's Note

This book is not a detailed analysis of the various cryptids and their respective merits. Loren Coleman and Karl Shuker have already covered that territory well. This book is about how I felt and I reacted to the situations I found myself in while hunting for proof of the existence of animals unrecognized by science. Some of my adventures were exciting, some amusing, some highly dangerous, but all are memorable. I took risks, and I am glad I did. I hope you enjoy it. But most of all I hope it inspires others, as time is running out for these creatures.

Chapter One

The Dream

It all happened in a flash. One minute I was frustrated, working in a lackluster job in a call center, the next, I was catapulted to instant fame with an interview in the *Los Angeles Times* and appearing on more TV shows than I could handle. I think it all began with a dream. It sounds cheesy, but I know now how big a dream it was.

When I was a young lad, like other young lads, I longed for adventure. I read with gusto about the daring journeys of David Livingstone, how he had searched the heart of Africa in the 19th century, and how he had seen things that most Westerners can only dream of. My mother would tell me of heroes like Chris Bonnington, the English mountaineer who made nineteen expeditions to the Himalayas, including four to Mount Everest and the first ascent of the south face of Annapurna. For me, Bonnington came across as the ultimate explorer: gentlemanly, modest, and charismatic, and of course, indisputably brave. I used to imagine being a Livingstone or Bonnington myself.

By the same token, ever since I was a small boy, I've been fascinated by tracking animals and observing them in the wild. My most vivid memories of childhood as a boy of seven was hunting for adders, Britain s only poisonous snake, on the cliffs of Cornwall, and being thrilled when we caught them basking in the sunlight. Closer to home, I was a regular visitor to the Peak District, and as I got older I was able to climb on to the moors and observe my first rare mountain hares. It was as a teenager that I first became interested in cryptozoology and avidly read about the Loch Ness Monster and Bigfoot. I never passed scorn on these accounts and I hoped these creatures were real.

Then one day my two passions came together while watching an episode of Arthur C. Clarke's *A Mysterious World*. The story focused on University of Chicago Professor Roy P. Mackal, and his hunt for a dinosaur in the Congo, the legendary *mokele-mbembe*. Mackal, the documentary stated, reckoned that a dinosaur had survived in the Congo. The area had been unchanged for thousands of years. It was remote and vicious; an almost impassable swampland surrounded the entrance to the beasts lair, a huge lake known as Tele embedded deep within the darkest region of the Congo. Straightaway, I was hooked. This was where I wanted to go. Was the animal real? I needed to find out for myself and that is how my adventures began.

There was one important point about Mackal that struck me: he didn't seem superhuman to me the way Bonnington was; he just had the guts to go for it. At university I studied history and I came to the conclusion that all great men are flawed in some way. That gave me comfort. I did not want heroes I couldn't measure up to. I wanted real men I could aspire to.

During this time I began to read about the expeditions of Sir Ranulph Fiennes, the British adventurer who was the first man to visit both the north and south poles by land and the first man to completely cross the Antarctic by foot. I'm amazed at how he was able to carry on despite the intense cold, and with his toes dropping off. I would never be able to measure up to him; I couldn't even walk in the man s shadow. The only ordeal I had faced was during drinking games with rugby and boxing clubs at university. Maybe Roy Mackal was one of a kind.

After university I got a dead-end job in telesales. It was dreadful work. Everything I did was monitored; even my bathroom breaks were recorded. But I was fortunate that I found it easy. I could dream of exotic foreign lands that I, Adam Davies, would one day travel to. The world may have already been conquered, but I was going to conquer it again for myself. So I traveled, spending all my money and my holidays seeing as many places as I could afford to see: from China to America, and from Egypt to Thailand. I went on safari to Kenya and the place took my breath away; seeing all those animals on the plains of Tsavo, I felt that I had entered the Garden of Eden.

Meanwhile, my unfulfilling career stretched on. Eventually I became a manager at another call center, and married shortly afterwards. Still my dream festered inside me.

Then one day I decided to act. I started to research the stories of the Congo. As it turned out, the legends about the *mokele-mbembe*, the horned beast of the Congo, stretched back to the 19th century and beyond. Early in the century, a German expedition had been launched to find it, more recently though, a Scottish man called Bill Gibbons had been there and returned convinced of its existence.

I also read of other modern day explorers and was encouraged by one in particular. Benedict Allen was no Sir Ranulph Fiennes. As a student, he had allegedly crossed the Amazon, making all sorts of mistakes and eating his dog in the process. He had no experience and seemed to tumble his way through with a mixture of luck and gal. He was my sort of guy.

I decided then that my balls were big enough to do it. I had tracked animals since I was a child as a hobby and was moderately fit and well traveled. What I needed was a team and some idea of survival skills.

I spoke openly about my dream, even in the dull environment of the call center. Although some people tried to ridicule me, it didn't bother me. What I wanted was to separate the wheat from the chaff and find people who were genuinely interested.

One person in particular seemed very enthused: a soft-spoken man named Jon. Although many people will say they would go on an expedition, most of it is idle pub talk. What impressed me most is that Jon did his own research on the subject. But that, as I would soon learn, was not enough.

So it was that Jon and I resolved to go on a survival course run by an ex-SAS man, named Mick. The Special Air Service is the principal Special Forces unit of the British Army. So Jon and I, and my friend, Antony Oldham, set off for the Brecon Beacons, a mountain range located in southeastern Wales that forms the central section of the Brecon Beacons National Park.

We were to meet Mick at a pub in south Wales, but we took a wrong turn on the way and ended up being an hour late. Mick and the rest of his lads were a bit worse for the wait. When I strolled in, I apologized to Mick and shook his hand. He was a small wiry man with eyes that twinkled brightly. After I bought a round of drinks, Mick and his team resumed their jollity, and he introduced me to the other members of the survival course and his two assistants: Dave, his son-in-law who was also an ex-marine, and Owen, a tough looking balding man in his fifties. There were six others in the survival group, most of them my age, and we quickly struck up a good conversation.

Mick proceeded to get completely plastered, as did his two instructors. When the pub closed, they were barely able to stand, so I ended up leading the party, with Mick giving me directions, to first camp. The next morning I awoke at first light to the sound of Dave puking his guts out in the tent next door. We then had breakfast and proceeded to learn survival skills as we trekked to our next camp, where each of the three teams would have to find materials to build their own structure for the night. As we walked Owen lectured on the various plants we could eat and how to prepare slugs. When I asked Owen how each of these delicacies tasted, the reply was always the same: "It tastes like shit."

During the afternoon we began to build our shelters and it became quickly apparent that Jon was just not up to it. Exhausted by the trek, he fell asleep leaving the rest of us to gather firewood and build the structure. Antony was furious and as we were all hungry by now, declared that he was going to have him for dinner that night unless he got a move on. To pre-empt this murder, I shook Jon awake, whereupon he made half an effort to collect some twigs. After we had built our structures, Mick came to inspect them. Our Bedouin construction passed his test.

We then learned how to skin eels, what kind of berries and plants to eat, how to trap birds, etc. During the rabbit skinning demonstration, Mick quickly dispatched the beast and then tossed its carcass aside after first squeezing the turds out of its body. Steve, a cockney who stood next to me, turned white; I thought he was going to throw up. Eventually he mumbled something to me about Mick being a fucking lunatic before turning away. I skinned the animals while Antony and Steve built the fire and the others prepared the wild mushrooms and peeled some potatoes.

Later on we lit a fire. Everyone had brought booze and we all got completely plastered. We ended up dancing around the fire as in *Lord of the Flies*.

The next few days rolled by, culminating in a fast stream river crossing.

As we limped back to our car on the last day, Antony pulled me to one side and said: "You can t take him into the jungle, mate, you know that." He meant Jon, of course.

I knew that, and so did Jon. At work on Monday, Jon came over to me and said: "You know how I was in the Brecons?"

"Yes, I know," I said.

"Truth is, I just don t think it's for me," he said.

"But you could learn!" I insisted, desperate for him not to buckle.

"Listen, Adam," he said, "we both know I'm not going to be up to it."

I had to admit he was right. Well, at least I had prevented cannibalism in the Brecons, and better to learn about an unsuited team member now than later in the field. I would have to find somebody else.

Chapter Two

War In The Congo

His name was Andy Sanderson. Andy worked the phones with me at the call center and during a break a couple of days after the survival course, he approached me.

"I've been hearing all about this expedition, this dinosaur trip," said Andy, "and I'd really like to go."

I was impressed by his knowledge as he had already researched the Mackal expedition, knew about the legends going back to the nineteenth century, and quoted reports by Herman Regusters who had been there in the mid-1980s. I was most impressed.

So together we began to plan the trip. We researched the kit we would need and began saving our money. I wrote to Bill Gibbons who kindly emailed me back with a few contacts and a few warnings about who to avoid. If I could get to a missionary station in Impfondo, he explained, there was a pastor there who through his contacts could aid me with the supplies necessary to get me through to Lake Tele. But first we would have to get to the Congo itself, and as there were no direct flights from Britain to Brazzaville at the time, we opted to fly Sabena to Kinshasa in the neighboring Democratic Republic. From Kinshasa, our destination would be just a short ferry ride across the Congo River.

When June rolled around, it was time to go. I put on my backpack and realized that it weighed a ton. *My god, I'm never going to be able to carry this bloody thing*, I thought. I said goodbye to my family and met Andy, wearing those ever-present sunglasses perched on top of his head, at Manchester Airport. We waved goodbye to Manchester, blissfully unaware of the dangers we were about to face in Kinshasa. Although I didn't know it at the time, Kinshasa was about to go into meltdown. A civil war was breaking out.

As we drifted about the airport in Brussels, our first stop on the way to Africa, Andy waxed on about the dangers of the jungle. "Apparently," he said, "if a leopard is about to attack, just before it does so, it coughs."

"Really?" I said.

"Honestly, mate, I was watching it on this wildlife program. We'll have to watch out for it."

Laughing at the thought of a coughing leopard, we climbed aboard our flight from Brussels to Kinshasa.

When the pilot announced that we were due to land at Kinshasa Airport in ten minutes, I looked out of the window. There were hardly any lights. Unlike most huge cities, Kinshasa was a city virtually without lights. I remarked on this to Andy and we both frowned.

As we touched down, the Sabena stewardess came over to us. "Do you know about customs?" she asked.

"Know about customs?" I repeated. "Yes," I said uncertainly, before continuing: "Please explain."

"You must be very careful in customs," she replied. I had read this in one of my books about Kinshasa, how it was likely to be a real effort, how it could take hours, and how bribes were commonplace. But I'd been to areas of the world like this before, and although I expected it to be a chore, I felt I could cope.

"Good luck then," she smiled.

To reinforce the proper behavior, I again re-briefed Andy: "Follow my lead, smile a lot, and be very deferential."

"Okay," he said, as we stepped off that tiny piece of Belgium, and into the sweltering heat of the night in Kinshasa.

But Kinshasa turned out to be unlike anything I'd ever experienced before, even other African airports like Mombassa. It was like walking into the Wild, Wild West. As we got off the plane, the atmosphere felt as thick as the steaming heat. Straightaway I saw two boys who couldn't have been more then thirteen years old standing against the airport building, all the while swinging Kalashnikovs by their sides. Other older soldiers eyed us and appeared edgy, their eyes darting from side to side.

We then had to go through the rigors of customs with three different aggressive officials. I responded in very bad pigeon French. Andy stayed quiet. Eventually we got to the last queue. We could tell this by the fact that there was an iron cage barrier in front of us, on the other side of which was a large group of Congolese, jabbering excitedly, some holding up signs.

In the queue, two little men in dirty white coats examined our paperwork. They looked like something out of the horror movie *Phantasm II*. Finally, they nodded at a soldier in customs and we entered the melee. And melee it was, as by now our bags had been brought off the luggage rack. However, they seemed to be up for grabs by everyone capable of climbing on the large pile of luggage. Andy had already scaled the luggage mountain and was busy wrestling with some guy who was either nicking his rucksack or trying to be Andy's porter, depending if you're an optimist or pessimist. Andy angrily pulled his bag away and we proceeded over to the search area.

A large woman who seemed to have brought enough pairs of shoes into the country to keep even Imelda Marcos happy was shouting at a customs official. He, in turn, was wagging a pink, stiletto-heeled shoe at her furiously. Bizarre.

Then it was our turn. A little man examined my rucksack first. As he did so I chatted about how much I was looking forward to seeing his country. Next to him stood a guy in plain clothes, just t-shirt and jeans, although both had designer labels. They looked through my kit but found nothing that interested them. Andy's rucksack was a different matter. He had a beautiful diving knife and the instant the guy with designer labels saw it, his eyes lit up.

"*Non, non, c'est dangereux.*"

"Dangerous? But I use it for cutting in the jungle," explained Andy.

"*Non*," he repeated.

"But it's mine," said Andy, by now raising his voice.

"*Non*," said the designer boy, puffing up his chest.

"Fucking hell!" shouted Andy.

At this point, the official clearly drew his finger across his neck making a *schlick* sound as he did so. Whether this was a genuine threat or not, it was time to intervene.

"Let it go mate," I said to Andy.

"But, Ad…"

"Let it go. You know there's nothing you can do about it. Forget it. Anyway, I hid my two knives in my food, so we're alright, we've got one each. C'mon, let s go."

He followed reluctantly.

It was about midnight when we got outside and the area around the airport was not a pretty sight. The pot-holed road that spread out in front of us was covered with a large crowd of men, all vying for business as taxi drivers. Although their vehicles might be past their best, their prices were way ahead of the times. As we tried to haggle, the flat fee turned out to be $100. There was no negotiation. If you tried to offer less, they simply refused to take you. Neither of us fancied the hike to our hotel, so I agreed to take the cab run by Francis and Michelle. I insisted my rucksack stay with me and began digging for my knife.

In the distance, I heard a gun go off. Andy mumbled "Shit," and Francis started chatting with me while Michelle drove. We talked about Kinshasa. He named places to go and we even discussed football, while I offered him cigarettes. I kept my hand on my knife, just in case though, and also suggested I might hire their taxi again so as to dissuade them should they decide to rob us. Andy, meanwhile, was taking in the city.

Kinshasa by night was indeed a fascinating place. Posters of the dictator, named Kabila, were everywhere. Such lights as there were flickered on and off continually. By the roadside, people huddled around small fires burning rubbish.

As the car moved slowly over the potholed road, we began to see increasing numbers of troops.

"What are you doing here?" asked Francis.

"Oh, we are tourists," I said. "We're not staying here for more than a couple of days, then we're off across to Brazzaville."

"It is bad here now," said Francis. "Kabila and the rebels." He shook his head slowly. "And foreigners, no popular. Americans, no popular. Belgians, no as well."

"We are okay though? We're English," I said. "And we have Manchester United, your favorite team."

"You are okay," he said, and laughed.

I had read in a guidebook that the area around the hotel was particularly dodgy, and indeed it proved to be so. As we approached, crawled along at five miles an hour, a gang of youths suddenly rushed towards the car.

Francis shouted, but as he did so, I raised my hand and brandished my knife, thrusting it in their direction. The lead youth was startled and stepped back. As he did so, Michelle drove out of the potholed section of the road and roared off.

"*Bon*, Adam, *bon, bon, bon*," said Francis, laughing. Michelle joined in and soon all four of us were roaring with laughter. Eventually, we arrived at the hotel and booked two rooms. I arranged to meet Francis the next morning.

"I'll probably need you to give us a lift to the ferry terminal tomorrow at least, if we decide to leave then," I said to him.

"Okay, *au revoir*, Adam and Andy," he said, whistling cheerily, no doubt happier with our one hundred bucks in his pocket.

We checked in at the hotel, both totally knackered.

"There's something seriously wrong here," I said.

"Too fucking right," said Andy.

It was not just the gunfire; there were crowds of people roaming around, and large numbers of soldiers. Even Francis told us it was bad. He said that people were now demanding better *cadeaux*, or bribes, for their services because they needed to squirrel away extra money for the bad times that lay ahead.

This was a serious worry to me. We did not have great reserve funds. The hotel was expensive; we could only afford to stay there a few nights. We had to get through to Brazzaville and up to the jungle fairly quickly, otherwise we would simply run out of money. It was a somber moment. I went to bed with a heavy heart, wondering what the next day would bring.

In the morning, Andy and I settled down to a nutrigrain bar for breakfast, and began planning the day's events. We found out that the hotel official who normally organized ferries and trips from the hotel would be able to see us at 10 am. I went down to see Francis and told him to come back at noon. We'd get a look at Kinshasa with him as our guide and probably go through to the ferry with him the following day.

As we spoke to the hotel official, though, it quickly became apparent that we were in big trouble. The situation here is very tense, he said. There was big trouble in Kinshasa. Apparently there had been demonstrations against foreigners the day before, and the rebels were on the city's doorsteps. The hotel official confirmed what Francis had told us the day before; prices had taken a huge hike and our "bribe money" wouldn't go very far. Though we might make it to Brazzaville, we wouldn't have enough money to get us back home. We thanked the man for his advice and went back to our rooms.

We spent two hours discussing our situation. The last thing we wanted to do was give up on our dream, but we had no choice. We had to leave. We simply didn't have enough money to pull it off in this volatile situation. It was a bitter lesson to learn. I would never again do an expedition without enough cash on hand to have the freedom to do what I wanted and to cover any contingency. Francis took us to the airline office and we changed our tickets. We would fly out tomorrow.

"I know it's clichéd, but I'm going to say it anyway," said Andy.

"So near, yet so far," we both said at the same time.

Our flight would take off that night. The city by now was beginning to implode, as expected. Crowds were openly on the street, and we hurried into the taxi. As we rattled through the street, we heard more gunfire, this time very close. Andy was filming.

"If they start shooting in our direction, keep filming," I said.

Suddenly, Michelle, who was driving the taxi, hit a pothole and shouted: "*Merde!*" He brought the vehicle to a halt. We had a flat. He kept saying *merde* as he frantically changed the tire.

Crowds gathered around us, and I decided we were better off outside the taxi than trapped within it. By the crowd's demeanor, it was apparent we were in serious trouble. As they came increasingly close and menacing, we began doing mock charges at the crowd to buy Michelle time. Finally, the tire was on, so we jumped back in the taxi and sped off.

Soldiers in large numbers were heading for the airport, and as we said our goodbyes to Michelle, one of them pointed a rocket launcher at my head, no doubt in a moment of bravado, in front of his mates. He demanded to know why I was leaving.

"*Mon pére est mort,*" I said. It was a quick excuse, and I apologize profusely to my father for saying he was dead, but it did the trick.

As we entered the gauntlet that is Kinshasa customs, we were asked the same question by an array of designer boys and uniformed officers. No one challenged my reply, which was good because we were already in a foul mood. I had to contain myself when an arrogant female customs officer made me drink my water to show there was nothing dodgy about it.

As we waited for the plane, there was an unusual group of people waiting with us: nuns, medical personnel, a large Belgian watching a mini TV set, and a barefoot American, clearly drunk and swigging on a bottle of beer. Two men came over to the Belgian and began shouting. He began shouting back, and they escorted him off somewhere. Then a very attractive woman came over and sat opposite us. Eventually she asked Andy to buy some shoes from her. Andy agreed to buy a few cakes from her, and shortly after we finished eating them, we were ready to board the plane.

We didn't speak much on the way back. We were both relieved and hugely disappointed at the same time. But when we arrived at Brussels, we began discussing what we were going to do next.

"I'm not going to give up," I said.

"Neither am I," said Andy.

"We need more cash though. We need to raise money, or save harder, or maybe even get company sponsorship," I said.

"Agreed."

As we arrived back in Manchester, Andy saw the baggage rack had a dinosaur motif on it.

"I don't believe it," he said. "Is that bitter irony or what?"

Chapter Three

World News

We faced the obvious ridicule when we got home. Then, two days later, when the United Nations began evacuating Kinshasa, people realized that we'd had no alternative but to return. Even though we had made the right decision, I was not going to let this setback defeat me.

At work I met a web-designer named John McDonald. John kindly agreed to design a website for me about our expedition to find the *mokele-mbembe* and soon we began getting emails from all over the world. John would pass them on to me, and I would reply in turn. One reply in particular interested me. It came from a Jan Sundberg.

Jan is Swedish cryptozoologist and from a look at his website, he seemed very active. He had mounted various expeditions, mostly in Scandinavia, and had recorded a documentary for the *Discovery Channel* on Selma, the Norwegian Nessie. I began corresponding with him after he expressed an interest in our Congo trip. He and Elle, his friend and colleague, wanted to come over to England to discuss the expedition with us. Andy agreed to put him up in his flat in Didsbury, and I, along with John, Andy, and my two friends, Antony and Keith, agreed to help entertain them while they were here. I was looking forward to meeting them. Jan seemed to be a veteran "monster-hunter." I could learn from him and would certainly volunteer for his next expedition.

As Andy went to meet Jan and Elle at the train station (they had come over by ferry), Keith and I drove in to meet them. On the way Keith slammed into the rear of another driver and as a result the front of his car was knackered; the car had a bent hood and smashed headlights. By the time we arrived at Andy's, he was considerably stressed.

"I need a drink. And quickly," sighed Keith – and Keith hardly drank.

"But first," I said, "we have to go and introduce ourselves to Jan and Elle, then whisk them off to the pub for lunch, where we have a photo shoot this afternoon." After Jan had first corresponded with us, he had emailed the *Manchester Evening News* to say he was coming over. As a result, they had dispatched a photographer to the pub where we were having lunch.

When we knocked on Andy's door, out popped Jan from behind Andy. Jan is a large man in every sense. He stands well over six feet, perhaps six-two or six-three. He is, however, quite large with a big belly. In contrast, his legs are quite slim and bandy. He has small eyes and large jowls. With a small baseball cap parked on top of his grey wispy hair, Jan, who is in his late fifties, reminded me of the captain of a merchant tanker. He gripped my hand firmly and introduced us to Elle, a much smaller, balding, bearded, and bespectacled man. We then all rushed off to the pub where we'd arranged to meet the photographer.

It seemed surreal doing a photo shoot. Jan, though, was the consummate professional and clearly loved every minute of it. When I baulked at the photographer's suggestion that I hold a plastic dinosaur, Jan was quite happy to do so, even suggesting (rather strangely) what angles would work best. Eventually, the shoot was over and the photographer told me the reporter would call me for an interview.

After exchanging a few pleasantries about Sweden, we began talking in earnest about the preparations for our next expedition for the *mokele-mbembe*. I explained what we hoped to achieve, and Elle, who was an acoustics expert and an extremely intelligent man, said he would research the possibilities, though he thought it would be difficult to construct a hydrophone for Lake Telle, as it was reputed to be quite shallow.

Jan stated that he would focus on the sponsorship side of things. "I have some excellent ideas about companies to contact, Adam. There really is some serious money to be made."

"Okay, Jan," I replied, "and I have some ideas for people to join the team as well." We needed to get moving, as the plan was to leave in a year's time.

"We shall make it, Adam, and it will be the best expedition ever!" said Jan enthusiastically.

We all raised our glasses to that; we had our beers, and Jan had his orange juice. Jan explained that he was a teetotaler; I would learn why later.

We had a great day of laughs and good cheer. In the evening, we went for a pizza, which both Jan and Elle hugely enjoyed, pizza being a bit of luxury in Scandinavia apparently. We picked up the tab to thank them for coming over.

The next day we all went for a hike in the Peak District, one of Britain's most popular National Parks. Jan, Elle, and I discussed some of Elle's technical developments while Jan spoke with Keith about his past expeditions, recounting the legend of the Seljord Serpent, the famous Norwegian Nessie. When we stopped for lunch, I reminded Jan of the need to be fit for the expedition.

"I know all this, Adam," he said, forcefully stabbing his finger in the air, "and I know that it will take me at least twelve months from today in order to get the fitness I need. I will see my doctor as soon as I get home."

John MacDonald looked at me and frowned.

We then joined in the conversation that Keith and Jan had been having about expeditions. "You know, on my last expeditions there were a lot of problems," said Jan.

"What sort of problems?"

"Well, we had a falling out," said Jan. There was a marine geologist on the expedition who was supposed to be an expert but apparently wasn't. "Well, the man was just a bloody idiot," ranted Jan, by now getting more animated by the minute. "He didn't have a fucking clue. Nor did the others." To be fair, at least Jan was honest about the fact that there had been problems. Apparently, the film company that had done a documentary on it had portrayed Jan and his expedition in a very poor light, not doing him justice at all.

The next day, Jan and Elle took the train for Newcastle, where they would meet the ferry to take them back to Gothenburg. From there it was overland to Motala where they both lived. Just before he left, Jan said: "I am planning Seljord for this summer in August. I will keep in touch about that as well."

"Please do," I said. I expected a fairly quiet run in to summer, but it would not be so, as it turned out.

I did a telephone interview with the *Manchester Evening News* reporter on Monday, and at the end of it, he remarked that the article would be appear in the newspaper by Thursday.

But no story appeared on Thursday. Nor on Friday. I phoned the reporter the following Monday.

"It'll be in, barring major events, of course," he repeated nonchalantly.

"Good," I said, "because I need some guarantees. *Maxim* magazine wants to do a feature on us as well, but we promised you the story first, so here I am."

The reporter paused: "Give me a few minutes and I'll call you back."

A colleague who was sitting next to me at the time said: "You're never going to pull that off. They're not going to go for it."

I smiled. "We'll see."

Two minutes later, the reporter called me back. "Okay, Adam, you're on. We'll do a double spread, full-color feature in the middle of the newspaper, Thursday."

"Great," I said, "and thank you."

"You have massive balls," said my colleague after the phone call. I burst out laughing.

On Thursday, the story appeared in the paper, as promised.

The day the *Manchester Evening News* published the article, it seemed that the world had gone mental. Everyone was phoning me, wanting my take on the story. Every national newspaper called me and interview offers came thick and fast. John and I agreed to do the *Big Breakfast*, a popular British TV show. We felt the publicity would help us get the sponsorship we wanted.

I think it's important that I restate my motives before I carry on. I was never in this business for the money. It was always about the adventure. Sponsorship for me was always a means to an end. It would be fantastic if I could do this for a living, but I was realistic enough to know this was unlikely to happen.

At first, we were a little nervous about all the publicity. Andy and I joked that they would portray us as two cocks because we were novices. However, we were honest about our inexperience and we sold the idea of our dream convincingly. I think our romantic notions plus the quirkiness of our enterprise worked quite well and was not badly received. Indeed, *Big Breakfast* turned out to be a huge laugh. John was extremely nervous and there was one tense moment when I thought he might freeze, but he pulled it back and answered the question. Keith and Andy ended up playing some daft game at the end called Tutti-Frutti. Keith, in particular, was delighted that he'd made it onto national TV as a strawberry. Previously his only other experience of being on TV had been on *Cheggars Plays Pop* as a child. "But I was disqualified for cheating," he explained. "The only time it ever happened in the history of the show."

"You're joking!" I said.

"No," he said. "The school was disgraced. The headmaster never forgave me."

"Well, mate," I said, "dressed as that strawberry, I think you can say you've redeemed yourself."

Jan started to email us with a list of the potential team members, including a herpetologist, an American SWAT team member named Scott Covey, and Eric Joye, the Belgian cryptozoologist. He also said his colleague, Inge Falk, who was a Swedish Forces veteran and an expert mechanic, was also interested in coming.

Even as John and I continued to do more interviews, however, the relationships between the team members were starting to deteriorate, especially between Jan and John. John was responsible for running the website, and as such would receive emails and then send them on to me for replies. However, we were both working full time as well and could not deal with the incoming email immediately every hour of every day. Jan was high maintenance and expected constant replies or acknowledgements, even if he was just making a statement. The situation became a little abrasive.

By now, Jan had put the finishing touches on his trip to Norway, and invited Andy, Keith, and me to join him. John couldn't go because of work commitments, but it seemed an ideal time for me. Not only would I get to meet most of the team members, except the Americans who I'd arrange to meet later, but it would be good to speak to Jan face-to-face and see if we could resolve our conflicts. It would also just be great fun!

To this end, I asked Jan if my friend Stuart, who was living in Finland, could come as well. Jan was eager to accept him. I would head the British contingent; the rest of the team would include Inge, Elle, and Eric, who were all set to go to the Congo as well. We would also be joined by Jan's number two man, Peter Falk. And a Swedish ultralight pilot would join us in the second week. I felt that whatever happened, my time with Jan was unlikely to be dull.

And so in August we set off from Manchester. In the middle of a seemingly endless forest lies Oslo. We were to spend a few days there before heading off into the Norwegian countryside and Lake Seljord. I found Oslo to be a small, compact, picturesque city. In the center of the city is the harbor. After looking around and visiting the Viking Museum, Keith said: "I can't get over how hot it is."

"It only snows in winter man," said Andy.

"Let's have a heavy night tonight" said Stuart. "I've got some Finish vodka. I'll show you how to snort it."

"Snort it! You're joking?" I said.

"No, I'm not," said Stuart. "With high drinks prices, it'll warm us all up before we head off to the bars."

"Fair enough" we all agreed.

That night at the hotel I had my first and last experience of snorting vodka. All I can say is you should definitely not do it in front of someone you fancy. As the vodka enters one nostril, snot seems to come out of the other in almost equal measures – a truly remarkable physical response. Afterwards, off we went to the bars, all nicely fuelled.

The bars were lively, and the people were all friendly. If you have the money, it is a good night out. At one point we were told we were in "the hardest bar in Norway." It seemed perfectly safe to me, and we had our photo taken with two girls who I swear said that they had "come down from the mountains" for the evening. They didn't look that bad to me.

We eventually ended up in a dance bar at two in the morning. We'd all had an excellent time, apart from Stuart, who was now hitting on any girl he saw in increasingly vain attempts, aware that the clock was ticking against him. When we decided it was time to leave, Stuart protested vehemently.

"No, it's definitely time, we need some shut eye before we head off tomorrow," I insisted. But he still managed to drag us into two more bars on the way home. We ended up leaving him, trying to convince two girls outside our hotel to come upstairs with him.

As I lay in bed I began to think of the last conversation I'd had with John. "What do you realistically think the chances of you finding anything are?" he asked me.

"Small," I replied.

"How small? I know you've done your research, but it's not likely there's anything in there really, is there?" he asked.

"I'd say two hundred to one," I replied.

"Two hundred to one? Not much chance then." As I reflected on that, the door opened and in staggered Stuart. Alone.

"Lesbians!" he shouted. He then fell asleep, fully clothed, on top of his bed.

The following afternoon we all piled onto the coach bus that would take us to Seljord. Andy whipped out his video camera and said: "Let the adventure begin!"

Chapter Four

The Seljord Serpent

In 1750, Gunleik Verpe was rowing his boat across Lake Seljord. This would be no routine journey across the lake for Verpe, however, as his boat was attacked and overturned by what he described as a "huge sea-serpent." This was the first written account of a physical encounter with the Seljord Serpent, but the local people say that legends about the creature go back farther than that.

In 1786, a local named Hans Jacob Wille wrote a description of the creature: "It is very peculiar and one of the most poisonous of all. It moves on the surface like an eel, and some years ago it bit a man in his big toe while he was wading across the Laxhol River."

Over the centuries the stories haven't gone away. In 1880 Gunhild Bjorge was doing her laundry down by the shore when a long serpent came crawling out of the water. She attacked it with a stick, "repulsed by its appearance."

In August 1976, in the Norwegian newspaper *Aftenposten*, a reader named Walther Berg wrote that in August of 1963, having decided to camp by the lake, "I got up early and went for a walk to have a look at the weather, when I spotted the big animal down in the water. It was lying perpendicular to the shore and the road with its head towards land. (I was not able to see all of the head. The animal was just resting in the surface, but I was able to see some sort of shapes along its back looking like wave forms.) I would estimate the length of the animal to be 30 feet. It was a horrible thing to watch. I feel that I have to tell someone what I saw. I have never told anyone before."

There have been many descriptions of the Seljord Serpent, but while it varies in size, most accounts put it at around five to ten meters in length. It has been observed on land, with people describing seeing feet or flippers on the creature. Most describe its head as being in the shape of a horse, although others say it looks like that of a crocodile.

Looking into the accounts of this creature, I really had doubts about whether this creature exists or not. There are, in my opinion, three massive arguments against it. First, although the lake is not exactly crowded, there has been a sustainable population of towns around Lake Seljord for centuries. I find it difficult to believe that there could have been no tangible physical proof of its existence in all that time, no matter how elusive the creature. Second, how could a lake of this size and depth—it is roughly fifteen kilometers long and nearly two kilometers wide with a maximum depth of 150 meters – contain such a creature? The water had always been considered to have a low nutritional value. Small-scale fishing does take place on the lake, but how could the lake possibly support a population of large beasts? Third, and *most* pertinently, there was never a lake at all until about ten thousand years ago, when the land was completely covered with ice.

In short, I didn't have much hope of finding the Seljord Serpent.

We were the first to arrive at Seljord. There was no sign of Jan or the rest of the expedition members anywhere. Eventually, we found the cabin of the camp superintendent, a man called Rodger. He showed us a spot where we could pitch our tents. I was sharing with Andy, while Keith shared the tiniest two-man tent I have ever seen. "You'll be very cozy in that at night!" I said.

After pitching our tents, we decided to have a look around. Seljord is something of a paradise squeezed in between mountain peeks, which are covered by vibrant forest. At one end of the lake sits the village of Seljord, a small rural community whose peace is only occasionally interrupted by "monster hunters."

"Magnificent," I said.

"Yes, and home for now," said Andy.

Eventually Jan and his team arrived, complaining of being tired.

"I'll tell you what though, Adam," said Jan. "I have spoken with Rodger, and the boats are ready. We can go out tonight on the lake with searchlights and look for the creature."

"Okay, let's do it!" I declared. We were keen and enthusiastic to get started.

None of us had driven a speedboat before, and I was elected to drive our small boat first. Onto our boat we loaded a searchlight, and off we went our little engine *phut, phut, phutting* into the darkness. Jan's boat, which was covered, followed shortly afterwards.

We had had little by way of introductions to the rest of the team, but at this stage we were really unperturbed – we were just excited to get started. I remembered that on Jan's last expedition he had gone for around the clock surveillance, with a team doing an eight-hour night shift, followed by two day-shift teams. We very quickly began to wonder why.

If boat noise wasn't a deterrent enough to the creature's appearance, our lights certainly would be. Besides, there was absolutely no way we could record or photograph anything, even if you did see it. And it was raining hard.

"This is fucking useless," said Stuart.

"Yeah, I agree. I'm going to have to talk to Jan about this tomorrow morning," I said.

"Well, he'll be nice and dry in his covered boat," said Stuart sarcastically.

Tired from our trip, we decided to just retire for the night and deal with it in the morning. Our tents were pitched near the shore of the lake. As I lay in my sleeping bag, I could hear the waves lapping the shore. The first night of the hunt for the creature had been exciting but frustrating. I also began to question Jan's judgment – it was clearly nonsense to do a sweep of the lake at night, just as he had on his last expedition. I decided that the next morning I would put our proposal to Jan about what we planned to do by way of observation.

In the morning Jan had a limnologist's report for us to look at. I remembered that the documentary had stated that Seljord was a "dead lake" – that there were hardly any fish in it. But Jan had had his own report commissioned and this had shown that there were concentrations of fish around the lake. The assertion that it was a dead lake was apparently nonsense.

I then overlaid the map of fish concentrations with a map of the locations where the creature had been most frequently sighted.

My plan was relatively simple. What I proposed to do was this: first, I would traverse the lake with my team to get familiar with its features and layout. During that traverse we would make a note of likely observation points. I felt it was pointless to toddle up and down the lake. I tried to get inside the mind of the creatures. They had to be naturally shy and would have learned to avoid mechanical noises long ago. The best hope we had was to drift, observe, and wait. We would do this in areas where there were large concentrations of fish *and* sightings of the creature. We would also try different times of the day, but I was most interested in dusk and dawn, traditionally an active time for predators.

After breakfast, I put this plan to Jan. He seemed remarkably compliant. "Yes, Adam that is fine, I must get ready because there are many journalists coming today and tomorrow."

"It doesn't seem like he's all that bothered what we do," said Andy.

"Well, that's fine, because that suits me, mate. Leave him and we'll get on with it," I said.

By lunchtime, we had begun our traverse. Stuart and Andy also entered the "cave" where local legend said the creatures slept. Of course, it was empty, but it was a great laugh. Most of the jokes centered on offering Keith as a sacrifice to the creature should it actually be living there and not be too pleased to see us. Keith was not totally comfortable on the water at this stage, and I reflected that he would have made a pretty sickly looking sacrifice.

"Weren't all the sacrifices for monsters supposed to be young virgins?" I joked.

"Well, that's why we're sacrificing Keith!" said Andy.

"I've always wanted you to call me luscious," said Keith by way of retort.

By now, it was near dusk and time to head back to our camp. When we returned, we found that Inge had erected two large pieces of equipment. One was an infrared viewer with telescopic lens set on the shore facing the lake. The other was a huge cylindrical object. I asked him what it was.

"It's a camera from the U2 spy plane," he replied. "It can take a photograph with high resolution up to one mile away."

"Fantastic!" I said.

"The plates are being fixed and are on their way," said Inge. "It will be working shortly."

All the Brits were impressed by this piece of equipment. But Keith was particularly interested in Elle's hydrophone technology. During the day, Elle, with the help of Inge, had erected two large masts on top of his caravan where he had stored his recording and monitoring equipment. Andy took Elle's hydrophone out in the boat that night, placing it in a deep channel that he determined would give him the necessary acoustic access to the lake.

That night we spoke briefly to the two other team members, Peter and Inge, but I had a long discussion with Eric Joye, the Belgian cryptozoologist and a wonderfully eccentric figure. He is one of the few people in the world who actually makes a living doing cryptozoology, mostly by lecturing on it, I assume. He also turned out to be a Belgian kickboxing champion, which brings up the obvious Jean-Claude Van Damme comparison. As I'd done some Thai Boxing in the past, I offered to spar with him at some point and he agreed. He also began speaking to me a little in French.

"You must practice, Adam, for when we go to the Congo," he said.

"Agreed" I said. "I'll get better. Speak to me in French as often as you like."

Eric also had a very engaging laugh, a sort of high-pitched semi-squawk. He laughed a lot, and we all joined in, unable to stop as soon as he started.

The next morning we got up at 5 am, ready for first light. Andy roused himself fairly quickly; Stuart and Keith were a little more reluctant, but we all were ready to push off within half-an-hour. Though we always adopted a lighthearted approach to what we did, we ended up working at least ten hours a day every day throughout the expedition. After all, we had come all this way and spent our hard-earned money to do this, so it was important to give the expedition our best shot, no matter how unlikely the goal.

We *phutt-phutt-phutted* across the lake to one of the good observation posts. Then we would sit in virtual silence for an hour at a time before moving on. We all watched different directions, so that the whole 360 degrees could be covered. As I sat there, I reflected on what an eerie place the lake was, with the mist creeping slowly over the lake's calm waters.

Jan had told us stories about the creature overturning small boats – would it do the same to us? But at this stage I still felt there was little possibility of finding anything. As the day wore on, we moved about the lake, repeating the process of observation, waiting for an hour, then moving on to a new spot. We saw nothing of any consequence. We did become more practiced at identifying wakes, however, whose wave ripples are often mistaken for a creature, and we discussed them at length, hoping that it would make the difference if we were to spot anything remarkable in the lake. Eventually it came time to return to the camp.

As we landed, Jan came striding over to us. "Did you find anything?" he asked.

"No, nothing I'm afraid," I replied.

"Well, I have some good news for you because Elle has."

We couldn't believe it and went rushing over to his caravan. Elle greeted us with a huge beaming grin. I will always remember his face.

"Listen to this," he said, jabbing excitedly at his computer. Over the loudspeaker came a low grunting sound, followed by more of the same. Elle reckoned it was two animals calling to one another!

We had all been impressed with Elle's expertise when he met up with us in Britain but our feelings were enhanced when he now also began playing recordings of our own boat moving through the water. From its movements, he was able to determine that we had a fault with our propeller on the boat, something that had occurred only that morning when we'd snagged a rock during a low tide. He couldn't possibly have known about it, as he obviously wasn't with us.

The mood in the camp was euphoric. However, it was masked slightly when later Jan asked to have a word with me. It is important to note that before we left for Norway, the guys had all agreed that I would be the leader of the British team. This meant that if an issue needed to be resolved with Jan, I would have to do it. Apparently Jan was unhappy about us having a few drinks the night before. My view was that if people had worked hard during the day, I had no problem with them having a drink at night – in moderation. But Jan felt that it created a bad image. Since he was the expedition leader, I agreed to respect his rules, as did the others after a few moans and groans of displeasure.

But Elle was upset with Jan for another reason. It turned out that Jan had already contacted all the media channels about Elle's finding and several of them were already on their way. "I have not had time to cross-reference this, and already he is off talking to the press," huffed Elle. Jan, however, was indefatigable and bounced around excitedly as we talked about the journalists who would be coming tomorrow.

The next day we were up again at 5 am, but agreed to take a break in the afternoon when there would be some TV crews from Norway there to film the expedition. As we arrived back, we saw reporters talking to Jan. He was already playing back tapes of the "creature" and declaring it to be a major success.

Keith propped himself up near the maps, where one crew were doing some filming. He wasn't looking at anything in particular; he just wanted to be on TV. Meanwhile, Andy and I sat on the jetty feeling bored. I was chucking stones. "This is a waste of bloody time," I said, thoroughly annoyed.

"Yeah, I can't see why he wants us all to be sitting around here, when all we are doing is watching him being interviewed," replied Andy.

The rest of the team had started to grumble, too. "I wonder when Jan will do anything," said Stuart, "other than talk about presentations!"

We all laughed. "Let's ask Eric if he wants to join us," I suggested. "We'll go in two boats. One team can go and search towards Seljord in the morning, while the other goes in the opposite direction. Gradually we can come together in the middle. It could double our chances of success, and we can see if it affects the creatures or their movement in any way, as Elle monitors them."

We all still had our enthusiasm, and although we needed Elle's success to be verified, we were pleased with the way things were going, especially since out own expectations had been pitched very low from the beginning. On my initial briefing a few weeks before, I had announced that we were going to have "a damn good go" at finding very little. The others all agreed, though we still thought it very unlikely that we'd actually see anything. We were determined to try, nonetheless.

That evening back at camp we all huddled around Rodger's TV to watch the Norwegian TV spot. Of course we didn't understand a word of it, but there were lots of shots of Jan babbling to some young Norwegian female reporter. There was also a quick shot of Elle, and of course Keith standing by the maps. "You are a media whore!" I said to Keith.

At the end of the broadcast, Jan had two announcements. First, that the microlight pilot would be joining us the next day, and second, that we were all going for a special meal tomorrow night at the house where they had stayed on their previous expedition. By this time the news of the "serpent" in Seljord was really starting to break. Jan announced that the film clips were going to be on CNN. Jan was now clearly in his element.

The next morning, Jan briefed me on information about what he was proposing to do with the pilot. The plan was to fly over the lake and look for any movement in it, while I would be filming. I suggested that if we did spot anything, we could send the boats to that particular area.

"Jerker is the best microlight pilot in Sweden," said Jan.

"Excellent" I said.

"Yes, we'll need him to be because with the mountains around the lake, Jerker says the thermals will be very dangerous. I doubt you would be able to fly like this in England."

"We are lucky that Jerker is such a good pilot then," I said.

"Yes, we are, but we will have to look after Jerker when he gets here," said Jan.

"Why?" I asked.

"Well, because he lost the use of his legs in a previous microlight accident."

When I told the others about this, the information was greeted with slightly alarmed laughter.

"Shit, he's paralyzed from a previous accident, it's dangerous, and we're supposed to go up in this thing?" said Keith.

"Yes, if you want to that is," I said reassuringly. They all volunteered.

That afternoon Jerker flew in, swooping low, and waving as he came over the camp. He turned out to be a wonderfully flamboyant character: very extroverted, speaking little English, but cementing it all with lots of backslapping.

Andy and I were to go first. Inge helped me get into a flight suit. It was a perfect day, untroubled by the clouds and rain that had greeted us on our first few days in Seljord.

"Ready, okay?" said Jerker.

"Ready," I said, gripping my camera in one hand and the bar in front of me with the other. We set off, it seemed, almost vertically. The air rushed past my ears, the fields quickly disappeared from my view, and we were off towards the lake.

If there is such a thing as reincarnation, then I have no doubt that Jerker would have been a World War I fighter pilot. He had superb control of the machine. Swooping low over the lake was an incredible adrenalin buzz. All the time, Jerker was singing, and his singing got louder as we banked first right, then left towards the mountain, then banking off. I loved it. The thermals were strong, and we could lose or gain altitude very quickly within a few seconds. Jerker also loved showing off. As we approached the camp, and realizing that we were being filmed, he swooped so low that Peter Falk actually ducked!

Eventually, the time came for us to land. In typical flamboyant style, Jerker took us as close to the treetops as possible. It felt like they were clipping our undercarriage. I never felt unsafe with him though. He was a great pilot. Andy went up next and got some fantastic film of the lake.

When we got back to camp, Andy and I took one of the boats out, and I remarked on how shallow the lake around Seljord had seemed to be from the air. This was also an area with a high number of sightings and a large concentration of fish. We agreed to concentrate our efforts the following day on two areas in particular. One we called "The Tunnel" because it was near a road tunnel on one side of the lake, and of course the other was the lake on the Seljord side. I now felt that if we were going to see the creature, our best bet had to be in one of these two locations. Andy and I would go to the Seljord side tomorrow, while Stuart, Eric, and Keith would head for the tunnel area.

We all felt quite high after the day's adventure. While the observation work on the lake was extremely tedious, the microlight flight had been a great relief. That night, we set off for our meal at the house where the previous expedition had stayed. We were all really looking forward to it, as mostly we had been eating dehydrated rations since they were easy to carry in our rucksacks. But the particular charms of these rations were wearing a bit thin.

"Moose pizza!" shouted Jan.

Keith looked stunned.

As Jan went into the house, Keith collared me in the doorway. "You've gotta be bloody joking, moose pizza?" he pleaded.

"Make the most of it," I said.

As we entered the house, Keith's enthusiasm for the night's culinary delights were not increased for in front of us hung, like a Christmas decoration across the room, a huge flystrip completely covered with little dead black bodies.

"Will you look at *that!*" he said.

"Do you want to inspect the kitchens, Keith?" I said, practically choking with laughter.

As we all sat down, Jan raised his glass. "I toast to you all, on a successful expedition and more to come."

Keith, who was sitting next to me, said: "It's fucking apple juice!"

That didn't surprise me, knowing Jan's aversion to alcohol; I couldn't imagine him having wine, even at a toast. However, the moose pizza, I have to say, was delicious. The flystrip was soon forgotten, and the owner of the house and his family all made us feel very comfortable. Keith, however, ate very little that night.

Back at camp, we went over to Elle's caravan to say goodnight, when we found Jerker supping a bottle of Southern Comfort that Jan had confiscated when he told us not to drink. Jan looked embarrassed. Apparently Jerker liked a little nip before bedtime, and Jan had felt unable to refuse. Keith, whose Southern Comfort it was, didn't object to Jerker having some but was not happy that he'd not been asked, or that we were not allowed to drink. Nor was I. There should be no double standards. I spoke with Jan about this and from then on we decided that it was okay to drink in moderation. We also decided head off into Seljord the following night, since it was a Friday.

The clock was ticking: would we get to see the serpent before we left? I looked through the night vision goggles to the opposite side the lake. "I know you're out there – I've heard you," I whispered to myself.

"It is only a matter of time, Adam." It was Inge. He had heard me and was standing behind me, preparing to bunk down. He had taken to sleeping by the shore in a sleeping bag. "

"I hope so, Inge. We'll see."

The next day, the two boats pushed off; Andy and I towards Seljord, while Eric, Stuart, and Keith went towards the tunnel. Like the day before, it was very hot and calm. The sun was shining and we saw fish jumping out of the water. We moored our boat in a shallow channel near Seljord, where a local woman had supposedly seen the Seljord Serpent. Apparently she had been swimming when she felt a huge creature brush past her. Though an avid swimmer, she had refused to go back in the lake ever again.

The day wound slowly on, broken up by intermittent conversations about going into Seljord that night, something we were all really looking forward to. We were about to head off to see the others at the tunnel, when… it happened.

About fifteen or so yards from the boat I saw the surface of the water bubble up. Something then rose from the water and swam towards the deeper part of the lake. It was black and I could clearly see three humps as it moved across the water. I never saw its head.

As it rushed past us, I shouted Andy: "What the fuck is *that*?"

"I don't know, Adam, I don't know," said Andy, who by now had turned on the video camera.

"Get the boat started, and lets get after it!" I shouted. At that moment I was full of adrenaline. I swear, if we had caught up with it, I would've tried to jump on its back and ride it like some modern day Captain Ahab. But despite my frantic efforts to get the boat started, the engine just would not respond.

It took a whole five minutes to get the boat started but by then the creature had long gone.

"That was it, that was it!" said Andy.

"I know, and we saw it," I replied.

The creature had been unlike anything I had ever seen. Rather than swim horizontally, it swam in an undulating manner. We speculated that it had been chasing fish out of the shallows, but that was just a guess. What we needed to do was confirm that we had it on video. To say we'd seen it without any substantive evidence would be to invite ridicule.

We had an agonizing wait though. Our boat had completely let us down. Eventually, we managed to row to a private jetty and from there two Norwegians who had seen we had broken down kindly offered to tow us back to camp. Nobody was there to greet us. By now, they had come to accept that we were unlikely to see anything. We thanked the Norwegians and hurried back to our tent to play back the video.

Fantastic! We had got it on tape. It was only a few seconds worth, but a few seconds would be enough. We played it over and over again. "Look at this," I said to Andy. "It looks like it has barbs on its back, just like the drawings in the legends." We decided not to tell the rest of the party until Stuart and Keith had returned.

I went up to the camp office to ask Rodger about bars in Seljord. "There are two main bars, one in an hotel and another where all the locals go," he explained. "There is also a nightclub on weekends where you can drink till late."

I thanked him, but before going off to tell the others, I asked him if he had seen the serpent, or at the very least, believed in it.

"Oh, yes, I believe in it," he said. "I've seen it several times." He then went on to describe a large black creature swimming past his camp. "It is very strange, but equally very real."

Eventually the others returned to camp. It turned out that their boat had broken down as well. "We've got a brilliant story to tell you," Stuart said.

"Excellent," I said, "but let me tell you our story first." Thus we broke the news. There was much excitement and backslapping, and Eric and I ended up in a sparring session by the lake for some strange reason.

"You could've taken him, Adam," said Andy.

"Cheers, mate, but I think your opinion is a little clouded right now."

The others headed off to speak with Elle. Inge meanwhile stood at the lakeshore, shaking his head at the two knackered boats. I asked Peter where Jan was.

"He's over by the showerblock," Peter replied.

I went into the loo, and Jan was standing there having a piss. I stood next to him, and began to pee myself.

"Jan, I've found it," I said.

"What?" he said.

"I've found the serpent, Jan, and I have just seen and videoed it in the water."

"You are joking!" he said, and sprayed his shoes as he did so.

"No, I'm deadly serious," I replied. The obvious humor of this situation was not lost on me, but there was also something else worth noting. Jan had spent over thirty years looking for a creature and had never seen one. We had seen one within a week. We had been extremely lucky.

By now, everyone on the team had heard the news and they were all very pleased. Peter, who with Jan had organized the sale of the sound recordings of the creature, was also interested in the commercial value of our videotape. Later Jan would tell me that he had tax problems, so I understood why money had become an issue for him.

Of course, everyone wanted to see the video immediately, but as we did not have adapter leads to play it on the TV properly, we decided we'd send a copy of the video to Jan after we returned to the U.K. That night we decided to hide the video very carefully. There is little crime in Norway, but we did not want our video to mysteriously "disappear" during the night. That would've been just our luck.

As you might expect, we were in high spirits that night. We'd been successful, we'd worked hard, and it was Friday night. The team deserved a break and who was I to deny them one?

The first stop was the hotel. The bar was dead and the drinks were even more exorbitant than they had been in Oslo.

Stuart then offered a solution: "I might go back to camp and get more of the Finnish vodka."

"No!" we all shouted.

At that point a Norwegian man who had been drinking alone in the corner introduced himself. His name was Triahorn. As we began to chat, he told us about Seljord and the legends: "Local people are divided on whether it exists, myself, I think it does, but legends on it go back hundreds of years. There must be something in it.

"And how's your Swedish boss?" he said.

"Oh, he's fine," I replied.

After we'd all finished our drinks, he suggested we go to the other bar. "All the townspeople will be there tonight. In Norway, very few people go out drinking until at least Thursday. If you drink before then, people view you as an alcoholic. Spirits are impossible to buy in this country, you know. They try to restrict the alcohol supply to control alcoholism."

As we walked along, "Trigger," as we now called him, also told us about the rivalry between Norwegians and Swedes. He was roughly our age, in his late twenties to early thirties, and spoke excellent English. We got on really well.

The bar was cozy, with a wooden interior. It looked a bit like one of those "olde English" type constructions that are created for tourist in places like the Cotswolds. Obviously though, this was olde Norwegian!

Trigger bought us a round of whiskies, which was incredibly generous of him. They must have cost him a fortune. He then went off to speak to some of his friends in a different part of the bar. We arranged to meet at the club later. It wasn't long though, before we were joined by another character whose name was Hans.

Hans was younger than Trigger, say twenty-five. He looked exactly like our stereotype of a Scandinavian: blond, blue eyed, with a skinhead and a square jaw, but he had no steely persona. Every other word was "fuck." Maybe he was trying to impress us, but Andy and I soon began calling him "Fuckjord."

Hans had a pronounced swagger to his conversation. It took us ages to convince him that we were really in Seljord to look for the serpent. Why else would four lads from Manchester have been in his bar? We were quickly joined by some of his friends, including a chubby hairdresser who Keith seemed to take a particular interest in. Fuckjord eventually believed our story and offered us some of the local hooch at the end of the night. Despite the obvious anecdotes about potential blindness, we all enthusiastically accepted. After all, there were no blind people in the bar to counsel us against having any.

Keith then rejoined us and said: "I can't believe I missed seeing the serpent today, you lucky, lucky bastards."

"Yeah, I know, and we've got it on tape. We've achieved more on this expedition that we ever hoped to," I replied.

We all raised our glasses. "Cheers!"

"Yeah, but did you hear what happened to us with Eric?" said Stuart, who then told us that on the way back to camp, their boat had cut out on them. But the cause of all the mirth was the bizarre way in which Eric reacted to their situation.

Stuart had suggested that they look for a tow, but Eric had become very animated at this suggestion. "*Non, non,*" he had shouted after kicking the boat's engine. "We must head for one of the islands." There were several small islands on the lake.

"Why?" asked Stuart.

"Because we can find berries and twigs," he replied "and live off them until we are rescued."

"Fuck off!" said Stuart.

"Yeah, I'd rather have those dehydrated rations than live off berries waiting to be rescued," said Keith. "How ridiculous can you get?"

"Why not, boys?" I said. "You could have been the wild men of the lake. Then there would have been two legends in Seljord. You never know, after a few years, after the stories had got round, we could have set up an expedition to find you."

"That's exactly what I thought you'd say. And that's why we outvoted the stupid twat," said Stuart.

The image of Stuart, Keith, and Eric living wild on an island dressed in twigs and moss was certainly a hilarious one.

Shortly afterwards we headed off to the club, along with Trigger and Keith's chubby little hairdresser friend. As the music played and because we were in such a good mood, we got the whole club going on the dance floor. Towards the end of the evening, a girl came over and asked me to dance. She was extremely attractive with long dark hair and intense blue eyes. Dressed in black, she curved and swirled in front of me. After we'd danced for about ten minutes, talking occasionally, she gave me a kiss, smiled, and then walked out of the club.

At the end of the night, as we were saying our goodbyes, Trigger asked: "Where is that girl you were dancing with? She was gorgeous. Has she gone back to your camp?"

"She's not with us," I said. "Anyway, I thought you'd know her. You said you knew everyone in Seljord."

"I do, Adam, but I've never seen her before," he replied.

The next morning at 5 am the team was less than enthusiastic about joining me on the boat, not surprisingly. Except Andy, who like me, had tasted blood. I left the others to it. Nothing much happened that day though.

The next day was our last day, and we all decided to go out on the lake as a team. Jan also came with us for an hour, but we rowed him back to shore for lunch. By now, he had little to do, as the journalists had gone and we had decided to keep our tape "secret" until we had had it analyzed. There had also been some friction between Jan and Elle over the selling of the recordings. Elle had been concerned over the tax implications, but by the last day, everyone was on reasonably good terms.

Rodger and his wife came out and gave us a big hug. He was a great bloke and we thanked him for looking after us. As we shook hands with all the others, Inge spoke to me: "Adam, you must get Jan fit if he is to go to the Congo. He cannot go as he is."

"I know," I said. "We'll be in touch after I get back."

At the airport, we all filmed our final thoughts on the expedition and said our goodbyes.

We had been successful. We had the videotape and the sound recordings of the serpent analyzed by different experts. Dr. Auld von Soldal from the Marine Research Institute in Bergen analyzed the sounds. After cross-referencing them with known species, and checking them against potential mechanical sounds, she concluded: "It astounds me to say that this is a unique species." It's worth noting that she had conferred with her colleagues before coming out with such this revolutionary statement.

Jan said that the Smithsonian Institute also analyzed the sound recording and had concluded that it "remains the only irrefutable evidence of the existence of a serpent."

Saab Image Systems analyzed the videotape, and they concluded that, "It was not faked and indicated a large possibly unknown animal moving through the water."

We were vindicated.

Back home, John McDonald could hardly believe it. A story on the results appeared in the *Manchester Evening News* the following week.

"I know it's incredible, but we really did find it," I said. The important thing is that we don't look like nutters.

"If reputable scientists are coming out and saying this, we don't look like nutters or fraudsters and that should aid in getting sponsorship," said John.

"Exactly," I said.

I couldn't have been more enthusiastic as my dream of the Congo drew ever nearer. But things were shortly to take a turn for the worse.

Chapter Five

Congo Or Bust

The *mokele-mbembe* is the creature that began my fascination with the hunt for rare mystery beasts, and it remains my inspiration.

The term *mokele-mbembe* comes from the Likouala language and basically means "river stopper." Legend has it that the creature, which is said to resemble a sauropod dinosaur, is still living in the remote vast swampland forests of the Congo basin.

The first written descriptions of the creature date back to a 1776 book by Abbe Lievain Bonaventure, a French missionary to the Congo River region, who described having seen enormous footprints in the region.

By the nineteenth century the creature began to be described as a dinosaur. Written descriptions and accounts said it was about the size of an elephant or a hippo with a long neck and muscular tail.

In 1959 Pygmies living near Lake Tele reported killing a *mokele-mbembe* to stop it from interfering with their fishing. But all who ate its flesh are rumored to have died.

The 1970s and 1980s saw expeditions in search of the creature by James H. Powell and Roy P. Mackal. Among recent attempts to find the creature are those by Bill Gibbons and Scott Norman, who has operated in both the Congo and Cameroon. Dr. Marcellin Agnagna and the American Herman Regusters, who sadly has passed away, claim to have seen the creature, but none of their claims have ever been corroborated by any other witnesses or with any sort of scientific proof.

Many people are extremely skeptical of the *mokele-mbembe* being a real animal. The physical descriptions of the creature are often intermingled with superstitions or mythology, as I would later learn, which makes their accounts unreliable. There has been no forensic evidence of its existence that scientists could scrutinize, as far as I am aware.

Critics also argue that the rainforest was mapped during the 1800s and 1900s, mainly to identify its resources, and as a result, any existing population of such creatures would have been discovered. I reject this argument, not least because many animals, including quite large ones, have been and continue to be discovered today.

But two points argue strongly against this creature being a dinosaur. All of the herd animals known to inhabit the forest are geologically young animals, being evolved much later on the dinosaurs. Secondly, much of the Central African rainforest is thought to be young rainforest, having been dry savannah as recently as 20,000 years ago.

I don't think *mokele-mbembe* is a dinosaur. The facts don't bear scrutiny. But I do think that the reasonably consistent physical descriptions given of the creature, and the vast wilderness in which it may operate, means it is worth exploring, whether it exists or not.

Thus we started to organize our expedition to the Congo. We divided our duties. I would organize some of the research and fashion replies to email inquiries, John would run the website, and Andy would try to get sponsorship. But our relationship with Jan posed the greatest difficulty. As I mentioned previously, we were unable to respond to Jan's emails as quickly as he demanded. On top of that, Jan was doing nothing to get himself in shape for the expedition.

Despite these difficulties, Jan, John, and I did a few broadcasts, one for the BBC and others for smaller independent digital TV stations in the UK, and carried on with sporadic media appearances. He proved to be a hugely entertaining character, engaging us with songs before we went on air. During one appearance, the BBC World Service sprung a surprise on me.

"We have a recording of a dinosaur here, Adam, and we'd like you to identify it," said the presenter.

"But no one's heard a dinosaur," I said, "so it'll be a guess."

"Yes, but this is what experts reckon they would sound like. Please can you identify it?"

"Go on then," I agreed.

A loud roar came across the speaker. "A Tyrannosaurus Rex," I said.

"Male or female?" asked the presenter; clearly miffed I'd rumbled him.

"Now you're being picky," I replied.

While these asides were pleasant, we were still no closer to our goal.

Meanwhile, things were finally coming to a head with Jan. Jan, in one of his more volatile moods, published an attack on John on his website, along with a picture of John. This was not a situation I could tolerate and we decided to part company with Jan. Jan himself offered to drop out at this stage. I personally think he realized that the physical rigors of the Congo expedition would be beyond him. The team Jan had brought together via the internet had also quickly dropped out after the initial media furor. In retrospect, I now think that gathering a team via the internet would never work; problems of discipline and correct skill match would always be a problem.

I was then contacted by Terence Bellingham, an ex-Royal Marine, who said he was also launching an expedition to the Congo. He said he had a wealthy backer named Paul, who was a publisher of coffeetable publications. At last, I thought, an opportunity with somebody who was not only wealthy but who also shared our vision.

We arranged to meet him in a roundabout way. Andy and I had arranged to appear on a small TV program called *Raw TV*, a bizarre show in which we sat with an alien expert and two white witches who barely spoke. Needless to say, we had nothing in common with the others. I was more interested in getting a free train ride to London to meet Bellingham and his rich mate.

After the show, we met the two of them in a pub around the corner from the studio. Paul really was in love with himself, but I found him both amusing and entertaining. He turned out to be a good friend of James Hewitt, of Princess Diana fame, and said that Hewitt would be joining us. I had real reservations about Hewitt, as I regarded him as something of a disrespectful individual. But Terence himself seemed switched on. Most importantly he had contacts in the Congo. My previous experience had taught me that to get through to Tele, I would either need a lot of cash or serious contacts – preferably both. Since I had neither, I needed Terence and he knew it.

"Look, it's like this," said Terence, "you've got the publicity, we've got the money."

He was right. And Paul wanted to be the leader of the expedition. This was not a problem for me; I was more interested in going than taking any credit. There was no ego in it, for me; it was only the adventure that mattered.

But shortly after we hooked up with them, Terence informed us that Paul had dropped out. He was supposed to be paying Terence a retainer while Terence worked on launching the expedition fulltime, but Terence had not been paid. So Terence and John began focusing on the business opportunities and worked to form a company. Terence also hired an agent, which caused further divisions with the likes of Andy, who basically saw the raising of funds as a means to an end. Without a backer, and as the months drew on and the frustrations mounted, we began descending into petty squabbles again. I began to think I would never get to the Congo.

At about this time, Keith, Andy, and I were to be interviewed for article for the *Mail on Sunday* magazine about our success in Norway. It turned out to be the funniest interview of them all. The reporter, a soft-spoken, very cultured individual, whose voice reminded me of a 1950s newsreel, politely asked us about our previous experience. Andy and I rattled out previous experiences in remote parts of the world, and then the reporter turned to Keith.

"And Keith, what previous experience have you had?"

"Er, Sea Scouts," said a nervous Keith.

The guy looked at him like he'd just crawled from under a hole.

I let out a half laugh and said: "Come on, Keith, there's no need to be nervous," and we all laughed. To this day we still tease Keith about that comment.

Our best bet for sponsorship now seemed to be a potential TV deal. We were in negotiation with two parties: the BBC and a likeable independent named James Cutler. The BBC seemed most keen on pursuing our idea for the Congo and we all went down to Bristol to meet with them. In London, Terence, along with zoologist John Riley and I, also did some video clips for James Cutler. However, it did seem that we were going round in circles. I didn't give a damn about the TV; it was a means to an end for me, but to others it was higher on their priority list. I don't blame them; I was just as single minded as they were, but just in a different direction, in pursuit of my goal.

Our agent ended up squabbling with the BBC. Even James Cutler's efforts seemed to be coming to naught. It was then that I decided to draw a line in the sand. I felt I'd wasted enough of my time on the Congo expedition. I gave it until the end of the year, and got the others to agree to that as well. If nothing was going to happen, so be it, but I was going to move on with my life.

To cut a long story short, both film deals fell through for one reason or another. James Cutler had been unable to come up with the goods, and the BBC, after continuing to pump me for information and much filibustering, announced they would not be going either.

By now, I had enough personal savings to have another go at it anyway. But the team was undergoing an upheaval. Andy and John had decided to drop out, each for his own reason. Terence, meanwhile, had recruited another ex-Marine, Nick Iddon, to join us. Even more important, through Terence we now had some much-needed Congolese contacts, in particular, Jacques Mayambika, who promised to sort out the vital Congolese paperwork and join us on the expedition. But Terence was now working with peacekeepers in Kosovo and as a consequence informed us that he would not be going either.

So there would now be just three of us leaving from the UK: Dr. John Riley, Nick Iddon, and myself. John wanted to go to do crocodile research. Nick was in it for the adventure and felt the idea of a *mokele-mbembe* a little fanciful. Though the three of us hardly knew one another, I decided to go anyway. It was now or never, and most importantly, Jacques had agreed to join us in the Congo.

Just six weeks before we were due to leave for the Congo, Jan called out of the blue. It had been some seven months since I'd last spoken to him. He had a proposition for me: "Would you like to come to Loch Ness in a couple of week's time?" He was coming over with a hydrophone and wondered if I'd be interested in joining him.

How could I refuse?

Chapter Six

The Loch Ness Monster

You can probably count on one hand the number of people left in the Western World who have not heard of the Loch Ness Monster.

In 565 St. Columba encountered the monster on the River Ness and commanded it thus: "Thou shalt go nor further, nor touch the man; go back with all speed." Whether you are religious or not, nobody can doubt that there have been centuries of sightings of Nessie.

There have also been many well-documented expeditions and numerous technologies employed to find the creature, everything from sonar to submarines, yet there is little in the way of reliable scientific evidence that can be scrutinised. Robert Rines took a famous "flipper photograph" that aroused predictable controversy with some claiming that the photograph had been enhanced. Very capable people with considerable resources have also conducted the searches for Nessie, but to date without solid success.

Various explanations of the sightings had been offered – everything from logs floating in the water to ripples of water moving across the loch's surface. There have also been a number of notable hoaxes. But it is clear that some witnesses did see an animal.

One particular account that impressed me was that of a Benedictine monk who, on a clear day, described seeing, from his monastery perched on the edge of the loch, a large, grey humped animal moving across the lake for several minutes. He had no axe to grind, no money to make, and no reason to fantasize about such a creature. What had he seen?

Many theories have been put forward as to what Nessie might be. Many think it is a plesiosaur, a fish-eating dinosaur that ceased to exist millions of years ago. But I feel the plesiosaur theory to be very unlikely. There are just not enough nutrients in the waters of the loch to sustain such large creatures. Besides, the loch itself is only ten thousand years old and sonar recordings have failed to pick up animals the size of plesiosaurs swimming through it.

Having said all this, I admit I wanted it to be real. Who doesn't really? After my experience in Soljord, I was prepared to give it a go. So in October, Keith and I set off to join Jan, Goran Rajalaard, and another Swedish bloke at Loch Ness. I was very much looking forward to the trip. The Loch Ness monster is the most famous cryptid in the world and it's right on my own back door.

On the road up to Inverness, Keith and I decided to stop off at a Travelodge in Glasgow. After having something to eat, we trundled off to the pub. The pub itself was not in the prettiest quarter of the city and it was in urgent need of a refit. It was dark and smoky and the guy behind the bar wasn't keen to serve us; maybe he didn't like visitors. It did produce one comic moment though. A woman who I'd love to say was really attractive, but in reality had a few of her teeth missing, came over to our table and said to me: "You are beautiful!" (Well, I'm not, really.) I though Keith was going to choke on his beer. The barman kindly ushered her away.

The next day we arrived at Inverness. We'd arranged to meet at Caley Cruisers. They had agreed to sponsor Jan with a boat, *The Highland Commander*, for the week. The boat seemed adequate for our needs and we loaded the equipment on board. It turned out that Keith and I would be sleeping in the prow of the boat. It was exceedingly cramped, with two small beds toe to toe and one cupboard. There were two bedrooms where Goran and the other Swede slept, while the dining area would be converted into Jan's bedroom for the night. Although we joked about our sleeping arrangements, we couldn't really complain. We were the last ones in and it was free.

The next morning we set off for the jetty near the castle, which was to be our base. Jan had indicated that this trip was primarily to be a reconnaissance for a more detailed expedition that he was planning in April, but inevitably he had contacted the press to tell them he was coming. The little boat chugged along to the harbor and, as it did so, it gave me an excellent opportunity to admire the remarkable scenery. The loch itself is vast, stretching from Inverness to Foyers, and is surrounded by trees and small hills around each side. All around the loch runs a coastal road, with viewpoints to stop and possibly, if you are really, really lucky, get a view of the famous beast itself. After all, you wouldn't be the first.

Loch Ness is charismatic and broody and I felt its charm. There was no such feeling for Keith, however. On the drive up through the highlands to Inverness, he had started to feel queasy with a stomach bug he had caught in Manchester. By the time we got on the boat, he was really starting to suffer and went below decks.

I did some videoing of the Loch and spoke to Goran as best I could, for his English was very broken, as he steered the boat in.

"Goran is a real joker you know," said Jan. At this point, Goran gave out a little whinny sound, like a donkey. I looked at him and laughed. He laughed, too. We were not all laughing at the same thing, but everyone seemed to be enjoying themselves.

After our initial reconnaissance the first day, Jan declared that he would outline his plans over dinner in Inverness. They were disrupted slightly as his press people had proved difficult to contact. Before he left for Scotland, Jan had given out Keith's mobile phone number to the media. But there was no wireless reception from the jetty, which was near the village of Drumnadrochit. As soon as Jan realized it, he quickly arranged an alternative number via his mobile phone. I'd never seen him work so fast!

Over pizza – nothing else would do for the Swedes – Jan outlined his plan. He again reiterated that his trip was really a preparatory one for the larger, two-week expedition in April. However, the hydrophone we would be using would be exactly the same. Goran said that some of the components of the instrument, which had been borrowed from the Swedish military, were classified. Jan spent half his time trying to chat up our waitress.

"Women like expedition types, you know, Adam," he said to me.

Keith started to laugh. I poked him in the ribs.

"Really?" I said. "You must have had a lot of luck with the ladies then, eh, Jan?"

"Oh, yes, but not for a while now," he admitted, laughing. "Maybe my luck will change on this expedition."

"No, he'll still be wanking like normal," Keith whispered.

Luckily, at this point, I was looking away from Jan. "Just be quiet and eat your pizza," I said to Keith. "Just be glad its not moose."

Towards the end of the meal Jan whipped out some contracts for us to sign. Basically, they stated that if we found anything in the loch while on the expedition, we'd divide the profits up from sales of photography. But one clause in this contract proved to be highly controversial. This was the "no booze" clause. I felt that moderate drinking was a good compromise, but Jan was having none of it.

Goran was not happy either. "I work hard all year and if I want a few beers, then I should be allowed to have them in my spare time," he said.

The discussions got quite heated, as Jan has a strong aversion to alcohol. But Goran got his way, as Jan needed him to work on the hydrophone. We just decided to let the matter drop that night and simply abstained from having a beer.

There wasn't much for us to do to do on this expedition, so we decided we'd go and interview some local witnesses about Nessie. Keith was still feeling decidedly rough around the edges, so we said we'd wait till morning. As I mentioned earlier, conditions on the boat for us were pretty cramped. Keith and I were toe to toe, and condensation dripped from the little porthole above us, stirring us in our sleep. There was no end to Keith's illness and with Jan rising at five in the morning, as usual, Keith did not get a good night's sleep. As I went off for breakfast, Keith was still in his bunk.

At nine Reuters showed up to speak to Jan. Just as the reporter step over to introduce himself, Keith came bounding out of the boat, sprinting down its small metal ladder, and promptly threw his guts up over the side of the boat. Jan looked horrified. The man from Reuters looked astonished. I turned and looked at the castle. All Jan could see was my silhouette against the sunlight, shaking as my shoulders went up and down with laughter. "Sorry about that," said Keith, as he wiped his mouth in front of a purple-face Jan.

The press was fascinated by Jan's proposal to launch a "trap" for the creature during his expedition in April. The idea of the trap had aroused huge controversy with many people claiming it would be cruel to trap the creature. In reality, as I understood it, the "revolutionary" trap was really nothing more than a standard eel trap. All the same, both Keith and I were not particularly comfortable with the idea. Besides, we weren't going to be present when he used it in April. We left Jan talking to a woman from BBC Scotland about it. Goran stood on deck looking bored, and by lunchtime, with Keith feeling better, we decided to go off to Inverness.

Inverness itself is an attractive town. It feels rugged, perched in the Highlands with an angry North Sea battering its coast, and its architecture is impressive, with two beautiful bridges at its heart. The town center is clean with plenty of buzzing bars and restaurants that befit its size. Keith amused himself by chatting up a bank clerk, telling her he was on the expedition.

Later, we went down to Drumnadrochit, the tourist town, to speak to some of the locals. People seemed reluctant to say they had seen the monster themselves, but everybody was keen to recount a second-hand tale of somebody they knew who'd seen it. But one man, who looked like a tourist's dream all dressed up in wellies and tweed, recounted how he'd seen some humps swimming through the water. Nothing anyone said really sparked me, though I did enjoy talking to the locals. Many of the older people loved telling a good tale. The younger ones were a little more suspicious, wondering if you were going to tease them or if you were a weirdo.

That night we returned to the boat and found the hydrophone in the water. It was clear that this was Goran's pride and joy.

"Nothing can escape this device," he said. "If it's in the water, we'll hear it." After the success Elle had had in Seljord with his hydrophone, I had little doubt that what Goran said was true. He invited us to listen.

After a brief explanation in broken English, I was able to distinguish fish swimming as they grouped to feed. There was no hint of anything unusual. "Surely if there was a sustainable population of creatures living in the loch, we would pick them up pretty much straightaway, as Elle had," I asked.

"Not necessarily," said Goran. He was as doubtful as I was.

We also visited Adrian Shine's Loch Ness Museum and I resolved to meet Adrian. He and Jan had had a falling out the last time Jan had visited Loch Ness. Initially we wandered through his museum and looked at all the attempts over the years to prove or refute the monster's existence. One in particular caught my eye: a small submersible, which looked like a large steel football, that Adrian had used to look for the creature in the depths of the loch. The museum made it clear that Adrian didn't think the monster existed. He speculated that people had it confused with a sturgeon.

When we went for coffee with Adrian, the first thing he said was: "You're with Sunberg aren't you?"

"Yes, I know you and he have had difficulties, but I'm really glad you could meet us," I said. "I've wanted to meet you for some time." He relaxed a little.

"Sunberg said he thinks it's a plesiosaur."

"I don't," I said "but I'm interested in your opinion. I know you've spent so much time here looking for it."

"I've done other things with my life, you know." He seemed defensive.

"I'm sure you have," I said. "I think it's an eel."

"Possibly," he said.

"Fishermen have told me about large eels being caught here," I went on. "It would also explain the land sightings, and the humps of an aquatic creature moving through the water." We went on to discuss the ecosystem in the lock and the unlikelihood of the plesiosaur theory.

"Well, I've not *just* spent my time doing this, but please do have a further look around," he said after we shook hands. "Oh, and don't trust Sunberg – he said he saw a UFO here once, which is total nonsense."

After lunch, we decided to go see Steve Feltham, who has spent ten years in a caravan parked on a beach at Loch Ness, convinced that one day he would see the monster. By now I was totally convinced he was wasting his time, but his motivation – or should I say obsession – intrigued me. His caravan, which was parked up close to the shoreline, afforded the best uninterrupted view of the loch that I could see around the whole coastline. We knocked on the door. He wasn't there. We decided to troop off to the pub nearby, where we learned that he away for a couple of weeks, helping somebody move house.

We stayed for a drink in the pub – coffee, mind you. Behind the bar, there were sculptures that Steve was selling to make a living. Though he might be an eccentric, he seemed to be well accepted by the local community; they didn't see him as just a loony. At first, I'd felt quite sorry for him, considering his rather bleak existence, stuck in that caravan. However, it became apparent that he enjoyed, or seemed to enjoy, quite a pleasant, stress-free lifestyle, and in beautiful surroundings no less. So who's the fool? I was the one with the 70-hour workweek, doing a job I hated.

The water had been too choppy that day for the hydrophone and everyone seemed bored, apart from Jan, of course, who had been able to entertain himself by talking to journalists, who had either phoned or come to see him. That night, Jan mused on politics.

"You know, I don't trust the Russians."

"Oh, why is that, Jan?" I said.

"They are getting ready for war."

"How can that be? They've lost most of the Eastern Bloc and the West is on better terms with them now than we've ever been."

"It's a trick, Adam," he said, by now wagging his finger. "Just a trick – you'll see."

The following day we set off for Aberdeen. We were off to see a chap from Simrad, a company that traditionally provided underwater equipment to Nessie hunters. Driving into Aberdeen, I couldn't believe how grey it was, though it is said to often win the "Britain in Bloom" contest. I think I can understand why. Every house was the same dull hue of grey. That would make the perfect canvas. Add the flowers and the whole place must light up. We were there in winter though! We had a good look around Simrad; they were very helpful to us and we had lunch afterwards.

On the way back we drove through the small town of Keith. Keith himself insisted we stop for a look round. "*In my Kinda Town...*" A poor Frank Sinatra impression followed. We were nearing the end of the expedition with only a few days to go.

That night, we checked out Aleister Crowley's house, which is perched on the south shore overlooking the loch. Crowley is probably the most infamous black magician of all time – the self-styled "beast," occultist, and alleged "Demon Conjuror." His place is reputed to be haunted. So much so in fact, that when the heavy metal band Led Zeppelin bought it as a curiosity, they quickly abandoned it. Now it stands empty next to a graveyard on a hillside. Whether it's the subconscious, imagination, or something else, I could feel the hairs on my neck stand up as we approached the house on foot. It struck me at the time that the only monster ever to have lived at Loch Ness was Crowley.

We were back pretty early that night, as the *Big Breakfast* was due in the morning, and Jan didn't want us to be late. That's when our little jetty, which had been so quiet all week, was a flurry of activity. Mike McLean had arranged to do an outside broadcast with Jan and it was obviously a send-up. He did three interviews with Jan. He rounded it up with a trip on board the boat, chucking a haggis into the loch in the process to tempt the beast out. Keith and I largely stayed in the background.

During breaks in filming, I took an opportunity to talk to Gary Campbell, the president of the Loch Ness Monster Fan Club. I put my eel theory to Campbell and he didn't object. I don't think Campbell believes the monster really exists; he's a businessman and fairly keen to perpetuate the myth, as the monster is undoubtedly good for business. I don't blame him.

By 9 am the crew of the *Big Breakfast* had cleared off, and with the expedition coming to a close, we decided we'd take our little boat to Foyers for lunch. It was during that journey that one of the funniest stories I have ever heard took place. As we chugged down to Foyers, I recounted to Jan our trip to see Crowley's place.

"This place is very mysterious, Adam," he said. "There are many strange things here we can only begin to understand. Everything from UFO activity to strange forces."

"Strange forces? What do you mean?" I asked.

"Well, I have been attacked once, and in Foyers."

"Attacked?" I exclaimed. "By what?"

"By something…some strange force. I was walking in Foyers. I had these special tweed trousers made in Sweden. I was in a bookshop in Foyers, when I suddenly felt something claw against my leg. I didn't know what it was. There was nothing there. When I looked down, I saw that my pants were ripped to shreds! I had not caught them on anything and they were completely torn. When I got home I sent them back to the manufacturers. They

THE LOCH NESS MONSTER 49

couldn't work out how it happened, and when I told them about the psychic forces at Loch Ness, they were astounded."

"I bet they were," I said, holding my laughter. Keith and I quickly went on deck to join Goran who was driving the boat. "I bet that's the only thing that's torn at Jan's pants," I said to Keith.

Keith laughed and said, "Yeah, it probably realized what it had let itself in for and decided to leave his knickers alone." This carried on for at least an hour.

On the last night, we went clubbing in Inverness, but before that we went for a goodbye drink with Jan and the boys at a pub on the way. We left them chatting up two middle-aged women. I have to say, Inverness has a high proportion of attractive women. Everywhere I looked there were raven-haired beauties.

The next morning, we waved the others off and headed down to Manchester.

"I'll be in touch soon" said Jan. "I hope to see you alive again, Adam!" He had cracked numerous jokes about my prospects in the Congo, even to the extent that his contracts had also bound "my heirs and dependents." I have not seen him since.

A few weeks later, Jan announced, not surprisingly, that he had found some "strange information" using the hydrophone. And his expedition in April took place as scheduled, but we were not there. Though it was initially greeted with a huge media frenzy, it ultimately ended in ridicule when a White Witch called "Kevin" clashed with him at the jetty. Jan lost his temper and threatened to chuck Kevin in the water, which made him a laughing stock.

What of Jan then? Despite his faults, I have to admit that in a funny sort of way I really do like him. Life for most people can be very dull, but Jan refuses to live like that. He chases his dreams and entertains in the process. I wish there were more people like Jan in the world.

As for the Loch Ness Monster, once you define what it is, ultimately it loses its mystique. After all, how many Japanese tourists would enter a place with a sign that read: Welcome to the Big Grey Eel Visitor Center?

Chapter Seven

The Last Great Dinosaur Hunt

The Congo trip was now just days away. I went through my kit checklist: rucksack, food, poncho, head torch, mosquito net, hammock, medical kit, survival kit, water bottle(s), knife, etc. I packed my food into my son Oliver's small rucksack that he usually took to nursery with him. It had his name on it in bold letters and was an England football team bag. *This will serve as a good reminder of my family when I'm away*, I thought to myself.

I wrote some letters in case I didn't make it back: one to my wife Laura, one to my mother, and one to Oliver. It may seem dramatic, but I was going to a largely unexplored area within a politically unstable country, so I didn't feel foolish about doing it.

A couple of days before I was due to leave, I received a phone call from John. We hadn't spoken for months.

"I just phoned to wish you luck on the expedition," he said.

"Cheers, mate, that's really good of you," I said. "How's the job going?" He was now working for an internet service provider.

"Oh, it's going fine, really well. It's a weird set up though."

"What do you mean?"

"Well, there are three floors, and I've learnt that each floor has a different type of fruit on it."

"What?"

"Yes," he said, "on the first floor they have your average apples and bananas, whilst on my floor, I have mangoes, and kiwi fruits and stuff."

"Well, what about the top floor?"

"I haven't been up there yet," he said.

"Maybe in the fruit hierarchy, they'll have like pineapples and nectarines," I declared, by now laughing my head off at the fruit oligarchy. Aldous Huxley's *Brave New World* of fruit had been reborn.

"No, I don't think so," said John. "They're too common!"
"Well, what do you think they'll have?"
"Lychees, I imagine," said John, completely seriously.

The night before the expedition, I arranged to stay at Nick's house, so we could travel down to Heathrow together. The plan was to go from Paris on *Air Afrique* straight to Brazzaville. Both *Air France* and *Air Afrique* were now flying direct from Paris.

The train journey down to Nick's took forever, with constant delays, but eventually I arrived, had a curry, and lay on my little camper bed, contemplating what would happen over the coming month. It had taken a lot to get to this stage, a lot of kowtowing and many frustrations, but I was going. I thought about the team members with their different objectives: John's was to study crocodiles, Nick was coming just for the experience, and I was going to hunt the *mokele-mbembe*, of course. We also did not know one another at all and there were massive age gaps between us. John was in his fifties, Nick his early forties, and I had just turned thirty. This combination of different objectives and people who didn't know one another would cause alarm bells in any expedition. But these were people who were prepared to go, and I felt it was now or never. I began to think about the hardships in front of me: the slog, the unknowns, the food we would be eating, etc. *I hope this dinosaur is worth it. It had better bloody be there*, I said to myself as I dropped off to sleep.

The next day, we departed for the Congo. At Heathrow, Nick gave his wife a kiss and we were off. As chance would have it, we found ourselves sitting next to a small guy with ginger hair, who introduced himself as Richard. We got talking and it quickly transpired that Richard had worked in Point-Noire in the Congo. Point-Noire is a coastal town where most foreigners who work in the Congo end up, particularly those employed in the oil industry. The Congo has rich coastal oil deposits, and where there's oil there's money.

"The Congo's a fucking dangerous place," he said. "Everyone's full of AIDS. They have this cough; they call it the 'custard cough.' Just before you 'go,' they spit out balls of mucus. Don't go anywhere near it. In the war, before they got us out, two of the wives in our compound were raped. It's a fucking terrible place. Be careful not to be robbed or be killed! Good luck. Oh, and enjoy the manioc." Although he was clearly bullshitting to some extent, it was hardly marvelous and encouraging news.

With his warning words still ringing in our ears, we met John in Paris for the flight to the Congo. There was a healthy queue for drinks while we were waiting for our flight. But out spirits soon changed as the excitement of the trip ahead raised our mood, and bravado kicked in.

As we approached Brazzaville, I looked across at Kinshasa, and remembered my last short and violent experience of the Congo. *I really hope we don't end up with a huge hassle through customs*, I thought. This time, though, Jacques had smoothed the way for us. He had been there for two weeks by the time we arrived and had been arranging the paperwork for us. I knew this could be an arduous process; it had delayed Bill Gibbons for weeks and we couldn't afford to be there months. I had to get back to work, and we all wanted to be back for Christmas.

Jacques had told us to be on the lookout for a boy soldier of about fourteen when we got off the plane. When I saw a boy swinging his Kalashnikov towards me, I waved back. *Time to adjust*, I thought.

Of course, the usual plethora of officials was still there to question us. People from the West normally come for business and generally they leave "the suits" alone. "Tourists" like us were a rare sight. But thanks to Jacques we sped through customs relatively rapidly.

Outside the airport, I go my first look at Brazzaville with its mixture of colonial buildings, some looking tatty, some not, and new shops. There was the usual array of small stallholders and markets that stretched endlessly by the roadsides. This time, unlike my first experience in the Congo, I felt in no immediate danger.

We were greeted by Jacques who took us by taxis to our hotel, the Bougainvillea. It featured none of the luxury of the big western hotels, but it was cheaper, functional, served food, and would more than do for our purposes. We quickly unpacked and I checked that my kit had survived the journey in one piece. It had. *Good*, I thought. *That's the first milestone over then.*

Our plan was to stay in Brazzaville for as little time as possible, as we would have plenty of time to see it on the way back. Unnecessary delays would suck up our meager funds. Our plan was to fly from

Brazzaville to Impfondo in the north of the Congo. Bill Gibbons and John Riley had both stayed with missionaries up there. Then, from Impfondo, we would go by truck to Epena where, hopefully, we would be able to get a boat to take us to the nearest village to Tele. And there, we would hire the villagers to take us the reputed 56 kilometers through swamp jungle, to the lake itself. Clearly, there were lots of variables and only a month in which to get there and back.

The first thing we had to do was to change our money, which was all in U.S. dollars and easy to convert in the Congo. Since the big hotels charged a hefty premium for exchanging currency, we drove to a crumbled part of town to do the deed. This is the way everyone changes money in the Congo, all the locals anyway. We waited in the taxi while Jacques disappeared into the building with our cash. I got out of the taxi and had a smoke while we waited. People ambled past us nonchalantly with inquiring but not unfriendly eyes. Many of the women wore brightly colored clothes, vivid lilacs and blues, often with headscarves to match.

Men openly acknowledged us with a cheery *bonjour*. By now it was very hot and sticky, and with a thin film of sweat breaking across my back, I began to feel the humidity for the first time. *Shit, I'll be carrying a heavy pack in this soon*, I thought. Ten minutes later, Jacques re-emerged, triumphant, carrying a large black bin bag.

"There'd better be more than magic beans in that bag, Jacques," I joked.

Jacques laughed. "Do not worry, *mon amie*, it is all here."

Back at the hotel, we counted the money and totaled up the receipts that Jacques had incurred since he got there. Jacques explained that even with his contacts, it had been difficult to get all the paperwork we needed. "Often people say, come back tomorrow, and then you have to wait and wait. It is very frustrating, even for me, Adam," he said.

Basically, the Congolese run their government on a hybrid of the French model; the French being the former colonial power in the country. This means that you must present your paperwork and yourself in front of numerous officials, each one with responsibilities in different departments. Or with responsibility for governing a particular area, say like a local *prefait*, for example. You needed the complete assent of each one to proceed. Despite the logistics of the trip, a *non* from just one of these guys could spell the end of the expedition. It was as simple as that.

But that afternoon we were to see a government minister I was actually quite keen to see: Dr. Marcellin Agnana, the Minister for Forestries and Fisheries. Agnana was a key figure in my research. I had first read about him in Redmon O'Hanlon's account of his expedition to look for the *mokele-mbembe* entitled *No Mercy* in the U.S. edition and *Congo Journey* in the U.K. edition. I think it is fair to say that Redmon does not give an entirely flattering account of the man; he portrays Agnana as a slippery, yet intelligent, individual who is a ruthless self-promoter. But given the tough climate of the Congo, it didn't surprise me that he had become a self-promoter. This image had been reinforced by John Riley, who had traveled with Agnana on a previous trip to the Likoula region (though not Tele) some five or six years earlier. The two had not gotten along.

It's important to note how, in tropical regions, even supposedly minor complaints can quickly spiral out of control. While in Brazzaville, John Riley's colleague, a German named Fritz, had picked up a blister. Walking had aggravated the blister. By the time they had reached Likoula, his whole foot had exploded into a pussy mess. Health care is not the best in the Congo; one report I read said there was only one doctor per one hundred thousand people, the worst ratio in the world. When they finally to got Fritz to a doctor, the doctor, upon examination, immediately handled the sterile surgical blades John had brought, thereby rendering them not sterile. The situation got so bad that they eventually had to air evacuate Fritz back to Frankfurt, Germany. When doctors examined his by now seriously bloated foot, they found the infection resistant to twelve out of the thirteen antibiotics they could have used on it. According to John Riley, he and Agnana had clashed over Fritz's situation, and over the fact that John had not shown Agnana enough "respect."

But I consider Agnana the foremost authority on the dinosaur, having been on several trips to the area, and I was really looking forward to meeting him. His office was situated in the grounds of the *Parc Zooalogique* near our hotel. This had previously been a zoo, but during the war people who were starving had broken into the zoo and eaten the animals.

We entered the grounds and walked towards the offices, a small one-story building complex surrounded by palm trees and the like. First, we were shown into a room to meet Agnana's assistant, Mr. Bonohmie, who was seated behind a small mahogany table. A short man, no more than five feet five inches, Bonohmie was dressed in a brown suit, white shirt, and black tie. Speaking broken English, he confirmed that there would be no problem with us going to look for the *mokele-mbembe*, which in his opinion, definitely existed, and he wished us luck for our expedition. Bonohmie was extremely pleasant with us; I got the impression that he didn't get many guests.

We then produced our *cadeaux*, or gifts, to sweeten the deal. We had footballs, note paper, etc., mainly small things. We also explained to Bonohmie that one of the papers required us to pay eight hundred thousand CFAs for a cine film permit that we didn't need. Our official passes were marked as press passes. We had to get across the message that we did not have large funds. As private individuals, we did not have the same degree of resource as, say, the Japanese or previous expeditions that had come to see them. Bonohmie understood this, and nodded sagely.

We handed over our meager gifts and were told we would have to wait for Dr. Agnana. As we were to learn, this waiting around was part of the system. Important men liked to keep you waiting to emphasize their importance. It was what Jacques had told us on the way in, how he'd often been told to "come back tomorrow." After our interview with Bonohmie, we sat on the steps for over an hour, and at first it seemed that we would not see Agnana at all that day.

Then suddenly Agnana appeared. He was immaculately dressed in a crisp white shirt and a suit with braces. (That's suspenders for you Americans.) I noticed the gold watch on his wrist. A lean man, roughly six feet tall, he strode towards us and shook our hands. When he saw John, he seemed a little startled "Ah, John, I did not know you were coming," he said. I would understand the significance of the curious pause in his voice later. We were shown into another room and were proffered seats around the table. Agnana sat at the end of it with a laptop by his side. Jacques, Nick, John, and I sat on one side. On the other was Bonohmie and a woman who Jacques said was the wife of an important general.

On the wall above Agnana hung a portrait of Sassou, the president of the Congo who had been deposed but then emerged triumphant at the end of the civil war. Propped on the floor was a picture of the old president. Clearly, these portraits were interchangeable. I smiled to myself.

At first, Agnana tried to get us to take Jean Phillippe, a government official, along with us on the expedition. Again, we reiterated the fact that as private individuals, we didn't have the funds. I have to say that Agnana was quite reasonable and understood that we were not high rollers. The general's wife, however, argued vehemently against us. Agnana clearly had to pay her respect and listened to her demands.

Each time she opened her mouth, she would start with "Pastor Jacques....*cadeaux*!" and my heart would sink. Jacques looked worried, but he reinforced our position and eventually she calmed down. As a face saving compromise, we agreed that if we came again, we would bring better gifts, particularly a typewriter as Redmond O'Hanlon had done. We were off the hook. I breathed a sigh of relief and smiled at the general's wife. She smiled back and nodded.

Now it was time to turn to the real subject at hand. "Dr. Agnana, can you show me where you saw the *mokele-mbembe*?" I said.

"Of course," he said, as I pulled out a map of Tele. "It was here, at an old hunters' camp, a few kilometers from where you first arrive. I'll marked it on the map."

"Please, can you tell me about it?" I said.

"Yes, I was looking across the lake, when I saw it rise out of the water. It was brown with a long neck. As I went to photograph it though, I slipped and dropped my camera. Phoowh!" he said gesticulating wildly with his arms, "then it was gone!"

John then spoke to him about his crocodile research and handed him a paper that he had written on the subject. Agnana thanked John, but I noticed he never looked him in the eye.

Just as we were about to leave, Agnana motioned for me to wait. "You know, Adam, there is another lake, Makele, that is where the *mokele-mbembe* really lives."

"Where is that lake?" I asked.

"Near Tele, not far," he said. Then we shook hands and said our goodbyes.

After we left Dr. Agnana, we went off to the Congolese TV station to meet Anatole, one of Jacques' contacts. The TV station was the scene of fierce fighting during the war, and a wall outside was pitted with bullet holes. Nick pointed out the different caliber rounds that were lying on the ground: Uzis and AK47s. We picked up Anatole and his assistant and then headed off for lunch. Anatole was trying to launch a film project about a chimpanzee namd Gregoire, an old, old chimp, some fifty plus years old, in fact, who amazingly had survived the civil war. He now resided in a zoo in Pointe-Noire. Anaatole said that Gregoire did not have long to live and insisted that his story would make a great film. I promised Anatole that would speak to a film producer I know when I returned. It was all I could do.

That night, we decided to go into Brazzaville. We nodded to the soldiers guarding our hotel and set off on foot. It's just a few kilometers into the center of town, or more specifically, to the brightly lit section of it where ex-patriots and oil workers go to be entertained. The walk was a pleasant one. The locals would occasionally acknowledge us, but there was absolutely no danger from them. The area has a high concentration of police and soldiers, immaculately dressed with rifles at the ready and patrolling regularly in open topped trucks.

We chose a small pizza restaurant to eat at, although its sign was riddled with bullet holes. Sitting to our left, some five feet away, were three, fifty-something, French businessmen accompanied by four young Congolese women, who Jacques indicated were hookers.

Over dinner, while being serenaded by a sax player, the discussion turned to the firm buttocks of the Congolese girls. "As John has his crocodiles, and you have your dinosaur," announced Nick, fuelled by the local crocodile beers, "I think buttocks should be my project!"

"I'll drink to that!" I said.

"I wish this sax player would fuck off," said John.

Jacques laughed, probably at us all.

Later, I spoke to Jacques about the meeting with Agnana. He confirmed we nearly didn't make it because of the quality of our gifts. Luckily though, Agnana had been on our side and eventually even the general's wife had been placated. She had even given us a mobile phone to call her on should we need anything in Brazzaville! I began to feel like we were making some real progress and it was time for a decent night's sleep. The next day we were off to the rapids.

In the morning, before breakfast, I took another look at my kit, trying to strip it down to the bare essentials. Jacques went off to get our final *livrettes* and at 11 am we left for the rapids, which lie just outside the city.

On the way down, John told us a story about the last time he was here. Apparently, he'd been walking through the city center when he'd heard a commotion.

"*Voleur! Voleur!*" the crowd had shouted, as the thief, who had been captured by the time John got there, was struggling to get free.

"They literally chopped him to pieces," said John "and left him for dead." Rough justice.

Eventually, we arrived at the rapids. The mighty Congo River crashed against the rocks, which groaned in the process, before rushing angrily off towards the distant jungle. By the riverbank were *pirogues* and small groups of men. The men were breaking small rocks thrown up by the Congo – obviously selling them later as bulking materials. It looked like damn hard work.

Taking a *pirogue*, we rowed out to one of the small islands and viewed the scene. The Congo felt impressive and powerful from there.

"The way things are going though, with the rock breakers, there won't be much left here," I said to Jacques.

"I agree, Adam," said Jacques "but they are poor people, what else can they do?"

"No, I understand," I said.

After a brief row about the taxi fare, we decided to walk back to the hotel. This was a good opportunity for me to talk with Jacques about the local politics and Zairen culture and also to admire the market stalls. Congolese market stalls sell an amazing variety of foods, everything from bats to caterpillars. As we were walking along, Jacques drew me to a woman squatting by the roadside. She displayed her wares in three large bowls, two of green caterpillars and one of black caterpillars.

"Try some," said Jacques, smiling.

"But which one to plump for?" I decided I'd go for the black ones. I picked up a couple of wriggling morsels, and popped them in my mouth.

"Good?" asked Jacques.

"No, they're fucking foul!" I replied. "They taste really bitter." Maybe they're better flambéed. Or with French fries. Then again, maybe not.

The market stalls stretched forever along the road. I also noticed many rounds on the ground, some of which had fallen into the open drains by the stalls. The drains were full of mosquito larvae: malaria traps. Eventually we came across some huge colonial buildings. They were brightly decorated in orange, peach, and lime green and all had palm-fronted gardens. Ahead of us, Nick and John had spotted a sign for cold drinks on sale and were gesturing for us to go in. I moved to edge forward and felt Jacques grab my arm.

"There is one thing I should say to you, Adam," Jacques said. "Marcellin does not like John. He says he is a bad man and has no respect for him."

"John told me the story of how they clashed the last time he came here." I said. "I believe that, yes, he can be abrasive in his dealings with some people; he was right to be annoyed with Marcellin."

"Yes, but it is the 'abrasive' that annoys the Congolese," explained Jacques.

By now we had entered the courtyard and joined the others for a drink. It was no place to hang around, though, as the worst kind of biting fly emerged from its slumber in the sand and proceeded to take chunks out of our legs. Even *Deet* didn't shift them. On our return, we finalized the arrangements for our trip to the jungle and paid for our flights to Impfondo. I was glad that we were off to the jungle. Brazzaville was an expensive city and we'd see more of it on our way back from Tele anyway.

That night I got about three hours sleep, worrying about Jacques being up and ready to take us at six in the morning. I woke the others at 5:45, the last possible minute, as they had indicated. The taxis arrived on time and then we shot off to the airport, where Jacques hurriedly joined us at the last minute.

The flight up to Impfondo isn't conducted in luxury. An old, old Anatov sat on the tarmac waiting for us. First, though, we had to get aboard and this would prove more difficult than I had imagined. We were subjected to the usual unnecessary searches of our belongings. The guards asked for a bribe, more *cadeaux*, and although we consented, John flatly refused and began arguing with Jacques about it.

As we queued to check our names off with the official who was letting people on the plane, he suddenly stopped me. "*Non*," he said, and waved his arm in my face. Jacques looked concerned and began remonstrating with the man, but to no avail. It turned out that because we hadn't offered a bribe, they'd taken one of our names off the list, and that name happened to be mine. Potential dangers are one thing, but having gotten so near, and spent so much cash, only to be denied at the last minute, was almost too much to bear, especially after my first experience. As I chain-smoked and contemplated the scene, Jacques was trying desperately arguing on our behalf.

Everyone went through, except for us. I had by now resigned myself to the fact that I would not be going. All sorts of thoughts rushed through my mind. Basically thought I would have to wait in Brazzaville, and either try to get another flight, or get out and go home. Then, Jacques came over to me.

"Adam," he said, "you have my ticket. You go."

"Jacques," I said, "no, it's not your fault, I can't."

"You must, Adam, it's your dream, you must."

"But Jacques, I..."

"No!" said Jacques, this time more firmly. "You must, I insist. The ticket will be no problem for you." He looked me straight in the eye. I was not meant to refuse a third time.

"Thank you, Jacques."

"Yes," he said. "You travel as me. Go, now go!"

We quickly said our goodbyes to Jacques, boarded the plane, and left him standing at the airport. John apologized for his mistake. To this day, I am still very grateful to Jacques for the sacrifice that he made. If Jacques had not been there to sort out all our paperwork and parley with the officials, we would have never gotten out of Brazzaville.

On the plane, we ended up sitting behind the only other non-Africans on board the flight: a French woman in her late thirties to early forties and an extremely tall American. The American, who must have been at least six feet four inches, wore a baseball cap and sunglasses and chewed his gum furiously. He began talking to us, reluctantly at first, as we set off.

"What are you doing up in Impfondo?" I aksed him.

"I can't tell you that," he said in a deep southern drawl.

"Why, will you have to kill us after you've told us?" said Nick. The American laughed and turned to shake our hands. "I'm Bob."

It turned out he was an ex-Green Beret. Apparently the war in the Congo had now reached the Impfondo area. Refugees by their thousands had been pouring into the area from across the river and Bob's job was to "clear a path" for the UN. I presume that meant making it safe for their vehicles and personnel. This was no doubt a necessary job. Many of Kabila's troops were reputed to be reluctant fighters, and there would be plenty of desperate armed men ready to flee.

As the plane climbed, we suddenly became enveloped in thick white smoke. "The plane's leaking a cloud!" I exclaimed. In reality it was a coolant pipe that had broken, but it did lighten the mood. I also have to mention the loo. With splashes of red, green, and yellow paint thrown all over, it had to be the most psychedelic bathroom experience I've ever had. A truly a bizarre flight.

Eventually, we neared Impfondo, and the creaking jalopy they had the nerve to call an airplane motored in to land. As we flew low over the jungle, I saw for the first time the famous Congo Basin, the jungle I was to cross.

We knew that there was both a missionary station and a hotel in Impfondo. We decided to opt for the missionaries, if possible. Bob very kindly offered to give us a lift into town in his UN Jeep. Nick waited behind to ensure that our luggage was offloaded safely, while John and I went off to find the missionaries, some of whom John had met previously.

As the UN Jeep pulled up outside the missionaries, Bob waved us goodbye. "Good luck in the jungle y'all. You'll need it!!" He gave a little chuckle and drove off.

The missionary complex was very attractive. It is set in its own grounds with three large brick and wood houses. In one house lived Dr. Joe Harvey and his family, while David Ohlin, his wife Diane, and their son lived the second. The third house, toward the front of the complex, was a guesthouse. We hoped they didn't have visitors. Joe Harvey opened the door, offered us a glass of water, and bade us sit down.

We were tense after all the day's dramas but I felt extremely relaxed from the moment I walked in the door, maybe because Joe, as I shall call him from now on, was an extremely calm individual. The missionaries were extremely good people. Paul and his wife worked at converting the locals, Joe explained, while he worked at establishing his own hospital in Impfondo, which was desperately needed.

Joe looked like the mirror image of Ben Affleck, the American actor, with the same height, build, and polite mannerisms. He also had a great sense of humor and I instantly liked him. We explained that we were off to Tele, I to look for evidence of the *mokele-mbembe* and John to look for crocodiles.

"You are more than welcome to stay here for as long as you need to stay," he said. "I will also help you see the local *prefait* and help smooth the path with him."

We both thanked him profusely and then we headed off, this time in Joe's jeep, to pick up Nick. Nick had collected our bags by now – as well as a couple of officials. One, who introduced himself as Jean-Phillipe, promptly confiscated our passports. Another, Michelle, it turned out, was Dr. Agnana's right hand man in Impfondo, and would be accompanying us on the trip. But our passports would not be returned to us until we had met the *prefait*.

Since everything hinged on our visit to the *prefait*, I asked Joe what sort of man he was. I needed to be able to push the right buttons, and in this country, I'd learned that flattery works far better than obstinacy.

"Oh, he's a hard man," said Joe. "Very powerful, kind of like a local warlord. Of course, he's appointed by central government, but in this area, his word is law." Then Joe added, with a lower tone in his voice, "and he's been heavily involved in the war." Joe smiled; I nodded. "But forget all that, guys, because tonight you're invited to dinner at the Ohlin's house, that is, if you'd like to."

"We would!" we exclaimed in unison. Back in Brazzaville, and mainly because it was cheap, I'd been eating quite a bit of manioc. I hate manioc. Even thinking of it now makes me want to puke. It comes as a lump and has all the texture and taste of wallpaper paste. Its only redeeming feature is that it fills you up if you're hungry. It turned out that the missionaries had a cook named Serge. We would definitely not be having manioc that night.

We retired to the guesthouse to unpack our equipment and wash up before dinner. It felt great to be there. Nick called this place the homeliest home – a reference to *Lord of the Rings*, and I can understand why.

Once clean and reasonably tidy, we walked over to the Ohlins. Paul gave us a warm welcome and patted me heartily on the back. He had lived in Central Africa for many years and was comfortable with the surroundings. He was in his forties, and although tall and thin, he looked strong, like a former American football pro. His wife, Diane, also shook our hands warmly.

They didn't get many visitors, so they were keen to hear our stories. We were just as keen to hear theirs and over a delicious meal of chicken and rice, we discussed our next obstacles.

"Well, I can help you get the truck down to Epena," said Paul. "It comes by here the day after tomorrow. From there you can see the local *prefait* and get a boat down to Boha. From there it's a trek to Tele through the jungle."

"Do you know the area?" I asked.

"Never been there myself. Still, I'm sure you will be fine. Maybe you can bring that dinosaur back with you." Paul, it turned out, liked to stalk big game. Not a hobby I was comfortable with, but his relationship with the Pygmies was both special and unique. He went hunting with them. He had also built their first church and was converting them to Christianity. He spoke about them with great warmth.

"You know, they are incredibly musical people," Diane added. "I've been learning to play some of their instruments. In fact, they're coming over tomorrow for a practice lesson."

"I'd love to meet them, I'd be really interested to learn more about their culture," I said.

"Well, no problem, Adam. We're going to the church tomorrow, we can cycle down and you can come with us," said Paul.

"Fantastic!" I said.

Once the evening was over, we returned to our guest lodge. For the trip to the jungle, we'd each purchased a liter of whisky. As things were looking up, we decided to crack one open. Joe had asked me not to smoke in the lodge; I respected his wishes and went outside. Looking across the Umbangi River, I could see bright yellow stars shoot across the distant night sky. Nick came to join me.

"Look at that!" I said, pointed to the "stars."

"They're artillery shells, Adam!" replied Nick. It was then that we heard the drums. Large harmonic drums booming away. Boom! Boom! A methodical beat. They grew louder.

We decided to investigate by taking the main road into Impfondo. The night had fallen, and we picked our way cautiously to the source of the drums. We found people swaying and singing around a burning fire, but as we drew close, they disappeared into the night. They were not hostile, but neither were they particularly pleased to see us. The scented smoke drew thick around us. As we left, the drumbeats began again. People were suspicious here, though not impolite. War and death were everywhere, I would learn on my return from the jungle. Considering what they had been through, these people were incredibly welcoming.

The next morning, we had breakfast and I headed off to the Pygmy Village, while John and Nick went to finish off the paperwork we needed before meeting the *prefait* that night. As we cycled down, Diane told me about the dangers presented by the locale. They had all left during the war, and their complex was saved by the pastor of the local church. When a raiding party arrived from Brazzaville to loot the complex, the Pastor stood his ground defiantly. "You'll have to kill me before you go one step further!" he had shouted. They went no further. The next day, the *prefait* had got hold of the party and kicked them out of town.

"It does feel a bit like the Wild West here, like we're living on the edge of civilization," I said.

"That's because we are, Adam!" said Diane, laughing, before adding more seriously: "But now the firing is right on our doorstep. The shooting is now only a few hundred yards across the Umbangi and we worry about the kids."

"I'm not surprised."

"Yes, but we are leaving soon for the Central African Republic – we need to put them in schools. We do worry about how things are going to be when we've gone."

Once we arrived at the end of the road that led to the Pygmy Church, we began passing various small shanty-type houses, some made of wood with tin roofs, others more robust constructions of mud bricks covered with thatched grass roofs. We were now off our bikes and walking the remaining kilometer or so to the church, as the pathway became impassable through the thick mud. It was the first time the jungle had presented an obstacle for me.

As we walked, Paul told me about the delicate interrelations between the Pygmies and the locals who lived on either side of the Umbangi. "You know, Adam, it's just as much about educating the Bantu here as it is about the Pygmies."

"How's that, Paul?" I asked.

"Well, the local people have for a long time not regarded the Pygmies as people like them. To them, they came from the forest; they are akin to animals."

"I'm very surprised," I said.

"Yep, the tradition goes back a long way. The Pygmies will often go into the forest for them to do jobs like hunting and they'll be paid peanuts on their return, like menial labor. Even the Christians at our church at first had a problem with us regarding them as equals. Now they're gradually coming to accept it, to treat them with dignity."

Paul went on to explain the details of their interrelations and how he had lobbied to change things over the years. I have often baulked at the idea of people losing their cultural identity as they adopt western values, even Christianity, but here I had no doubt. The church had served a double purpose of bringing the Pygmies together and helping to end racial discrimination. I, too, wanted it to be a success. Apart from their animist beliefs, Paul and Diane assured me that the Pygmies' cultural traditions were being preserved.

Other than the Ohlins, it appeared that the Pygmies had had no other western visitors and I was an instant celebrity. They were incredibly warm and friendly and seemed delighted to see me. Both men and women ranged in sizes, some as small as four feet two inches with the tallest being five feet seven inches or so. Some of the women still had pointed teeth and facial tattoos, which were considered a sign of beauty, though many had abandoned this custom. All wore conventional clothes, albeit in varying states of repair.

Paul had learned Aka Pygmy and spoke to them in the dialect. After the usual conversations about where I came from, we strode into the church. It was a small red brick structure, no more than twenty feet long, with an angular tin roof supported by wooden beams. Brick seats arranged horizontally inside served as the pews. To the right of the church, were two small mud brick huts. I understood enough of their gestures to know that the Pygmies were clearly very proud of their church.

At the far end of the church were three large drums. I strode through the church and did a turn on them – I was no Keith Moon, but my audience loved it, laughing and clapping the whole time. I earned my first ever-standing ovation. Several people even swirled to my thrashing of the drums. Paul looked like he was about to expire with laughter.

"Well congratulations, Adam," said Paul. "I reckon you're the first to hold your own 'concert' with these guys!"

When began Paul preaching, a little guy with a stick patrolled down the pews, encouraging the audience to make noises in the right places. Paul and Diane had been encouraging the Pygmies to read and write and everyone proudly produced their books. The guy with the stick really wasn't needed, though, as it was clear that everyone was entranced. It was then that I realized just how musical these people were. To the beat of the drums, beautiful melodious song burst from the swaying Pygmies. It filled the church, seemingly straining at the roof. I was entranced.

At the end of the service, I asked Paul if we could talk about the *mokele-mbembe*. One young man spoke up.

"I have heard it roar, it comes from the water with a *whoosh*!" he said, throwing his arms up wildly.

"It is very dangerous, it will shoot you with lightning from its eyes," said another.

My heart sank. "I don't think I'm going to get anywhere here, Paul."

"Wait, Adam, speak to this guy, he's respected." An old Pygmy shuffled towards me. "He says his people rarely go to that area now, as it is considered a sacred animal. Only a very few have seen it."

"Has he seen it?" I asked excitedly.

"Yes," came the reply. "He has seen it twice."

"Ask him what it looks like," I said. The old man drew a picture of a Brontosaurus-like creature on the ground. Unusually, it had a horn on its head.

"It is aggressive, it will attack. He says be careful. That is, if it allows you to see it!" said Paul, translating for the old Pygmy. "It lives in the forest, not the water, though it can travel through it. It eats plants. He says that's all."

"No one else has ever mentioned the horn before, in the books I've read about others who've come here," I said.

"Adam, not all those who've written about this place, have actually been!" said Paul, and winked.

I waved goodbye to the Pygmies and returned to the complex.

With the remaining paperwork sorted out, we were now ready for our meeting with the *prefait*. Joe had to see the *prefait* about expansion plans for his hospital so it was an opportunity to kill two birds with one stone. Meeting the *prefait* was like something out of the film *Wild Geese*; if ever there was a stereotype of an African Warlord, this guy was it.

We pulled the jeep into a muddy courtyard where a sentry box with two armed guards stood. Joe asked us to wait; he walked towards a massive, several story stone house. After a couple of minutes, he motioned us forward and we were shown into a huge courtyard that formed the central part of the warlord's villa. Within it, were four or five gleaming, four-wheel-drive trucks.

A woman, dressed much like the general's wife in Brazzaville, came out to greet us. As she sat next to us, we began to talk in a loud voice about what an honor it was to be there. All of us nodded in agreement, over emphasizing our delight. After about half-an-hour, she suddenly rose and motioned us forward.

We were shown into a large, executive-style boardroom that was plushly decorated. Large renaissance-style pictures hung on the walls. In the center of the room was a vast mahogany table. At one end stood the *prefait*, measuring about five feet seven inches and dressed in a khaki colored suit. He looked muscular; I remember, in particular, his thick neck, like a bull's. His eyes were small and darted between us in quick succession.

Once Joe introduced us, we all smiled and sat down.

"Why do you want to go to Tele?" he asked me.

"I would like to find the dinosaur, the *mokele-mbembe*!" I replied in a deferential manner.

"*Mokele-mbembe*!" he said chortling and jabbing his finger at the papers. "That is good. It is real you know!"

"Yes, sir," I said, "and I intend to find it."

"Good, good. You have my permission."

"Thank you, sir, I am very grateful."

"Good."

We briefly explained our backgrounds and then Joe motioned us to leave. This was a huge relief. While we were in Impfondo the *prefait* had the power of life and death over all of us. I doubt that we were ever in any danger, but one nod from him and that would have been it. Ten minutes later Joe was out, too.

As we drove back, I turned to Joe and said: "You know, Joe, I never asked his name!"

"His name? His name is …. Gilbert," said Joe. "He is a soldier and a porter."

"Good grief," I said. I'd lost it.

That night we again joined the missionaries for supper, and afterwards, I took the opportunity to talk to Paul about his faith. One question in particular troubled me.

"Paul," I said, "I know you're devout and that you've given your life to Christianity. But what about all the suffering that you see around you? How can you reconcile this with the idea of a loving God?"

"People often ask that question as an excuse for not believing or doubting, Adam," Paul replied.

"I understand that, Paul, but not in my case. I had a sister who died of cancer when she was only three; it was very painful for her. I'm asking because I want to understand."

Paul turned towards me. "I'm sorry, Adam. I don't have all the answers. I'll always have questions myself. I think God wants to intervene but it will break his covenant with man – it's almost like he can't. He does, however, want us to help, to help our fellow man. God loves everyone; he loves you and knows every detail about you. In the end faith comes down to trust, just as much as belief. It's not logical and cannot be defined. When you believe, you know."

After dinner, we made our way back to the guesthouse in the dark, always on the lookout for cobras in the grass. One had nearly bitten Paul a few weeks before.

The next morning, we were up at six and our team was there, ready. Our soldier wore a full combat uniform; he was about six feet three inches and slung his Kalashnikov casually over his back. He was holding four French stick loaves of bread. He wore sunglasses and a cigarette drooped from his top lip.

"Why the loaves?" I asked.

"I do not like the manioc," he said.

"I'm sure we'll get on well then," I replied with gusto.

Michelle, Agnana's deputy, was also there to see us off. He was to accompany us, but would follow us down to Epena the next day. Apparently, he had a message to contact Agnana that evening on the radiophone; there were no telephones here. The truck to Epena pulled up and we said our goodbyes to Paul. We pulled out our dollars out to pay him.

"Keep your money until you come back," he said. "You may need it! The last expedition that came this way was Japanese. They didn't keep track of their money so the villagers held them hostage. One Japanese girl came out from the jungle a few weeks later, which they'd let go. We had a hell of a time wiring money through to rescue them. Make sure it doesn't happen to you!"

The truck journey was something else. Nick and John got lucky and sat in the front with the driver. Philbert climbed on the top. I sat in the back with about twenty other people, a couple of soldiers, and bags of rice. You name it, I think it was in that truck.

On one side of me were two men, one of whom proceeded to pick what looked like a tick out of the other man's neck. But I had more pleasant and entertaining company on the other side, in the form of a woman and her monkey. The monkey, a pet, was fastened with a loose cord and spent most of its time jumping between her knee and mine. I love monkeys. I remember as a child seeing one in a pet shop while on holiday in Wales. I went to the pet shop every day to stare at it. My parents had to drag me away.

The journey took about three hours. We stopped a few times; once to pick up a red antelope and the second time for a porcupine, obviously someone's dinner later. The third time we stopped at a conservation center. Throughout the Epena area, there had been heavy flooding with the rains and the ground had almost been washed away. Finally, we arrived in Epena, which was very good news for me, as I had the numbest backside imaginable.

The local official down there was a Mr. Paco. He allowed us to sleep in some lodges he was redecorating for some World Wildlife Fund officials who would be arriving in a few weeks, about the time we returned from the jungle. At night we wandered off into town with Gilbert, our friendly soldier, proudly posturing in front of us, his "shades" neatly in place. A woman barefoot in a grass skirt wandered past us – the only time I ever saw any woman dressed like that. I purchased some peanut butter and some bread at a market stall and Gilbert and I settled down for the evening, leaving the manioc well alone.

Michelle was due that afternoon and was scheduled to radio Mr. Paco at 7:30 am to confirm his arrival time. We would then go by boat to Boha. However, the radio call did not go as planned.

"John Riley cannot go to Tele," said Paco.

"Why?" I asked.

"Michelle says that Agnana says he is forbidden to study the crocodiles. Agnana does not like Mr. Riley it seems."

"The feelings are mutual," said John, understandably.

"Michelle wants to speak to you, Adam," said Paco.

"Michelle, please don't stop us here, we've gone so far," I said.

"But what can I do, Adam? He has said no. If we allow it, I will lose my job."

"Well, what if he is allowed to go, but not to study the crocodiles," I suggested.

"I am not sure."

"Listen, Michelle," I went on, "you will be there, you can check on him and we can hunt for the *mokele-mbembe*."

"Okay, Adam, okay."

"Thank you, Michelle."

"Okay." Mr. Paco also consented to this plan.

I was grateful to Michelle, he'd stuck his neck out, but obviously John was gutted.

"You may still be able to do some research, mate," I said, trying to reassure him.

"Maybe," said John. There was little we could say at the time.

Michelle arrived and we loaded the motorized *pirogues* with food and fuel for the journey to Boha. As I climbed in, a man offered a boa constrictor for sale. "I've got nowhere to put it in the boat," I said.

He laughed. "Round your neck, *monsieur*."

"No, thanks," I replied. "I've just eaten."

The journey to Boha was quite an experience. The Likoula region is beautiful and unique. Our boat screeched through the marshland, dancing among the reeds, and all manner of beautiful birds flew around us, everything from eagles to parrots. Every now and again we would stop to untangle reeds caught in the motor or to move loose branches of semi-submerged trees that blocked our path.

Paco was obviously enjoying the cut and turn of the boat, but eventually it started to rain and we decided to stop off in the village of Etangi, where we became instant celebrities. All the children came out to greet us. They were happy, friendly kids and we chased them round, pretending to catch them, which they loved. I shared a couple of cigarettes with an old bloke who was busy weaving. It was a pleasant stop.

After the rain let up, we carried on to Boha. I could see the village from some distance, with *pirogues* parked on the shoreline.

We scrambled up the embankment and said our goodbyes to Paco, arranging when he would come back to get us. At last we had arrived.

The locals, although not unfriendly, did not show the same degree of interest in us as they had in Etangi or Epena. However, as it was nearly dusk, we went straight to pitch our tents. The Chief came out to meet us and showed us to a place on the side of his courtyard, the main square of the village opposite his own hut.

"We will celebrate tonight," he declared. I thought he was young for a chief, only in his early thirties, but I was to learn why later. I unpacked my gear and settled down.

The celebrations were to take place in the main courtyard. First, we presented our gifts to the Chief, which included a football and pump, cigarettes, etc. We were each offered chairs, and an old guy, who I would later learn was a notable person, brought us a drink.

"Adam, *pour vous, le whisky du village.*" I can say now that village whisky tastes nothing like single malt. It smells like toilet water after a curry, and it tastes, well, even when I was running the rugby clubs at university, I never did that for a bet. Anyway, like all true Englishmen abroad, we had several glasses of this potent stuff. A large fire was lit and drums were brought into the middle of the square. Everyone joined in; it was great fun now that the ice had been broken. The old guy who had given me my first village whisky gave me an ultimate compliment: "You are a very good dancer, Adam." What respect!

At the end of the night, another old guy showed us where the cesspool pit was. He must obviously have anticipated our reaction to village whisky as he held his guts and laughed when he showed us where it was. It was diagonally across from our tents, hidden by a small felt curtain.

When I woke up the next morning, it felt just like the party was still going on inside our heads. The old man was right about the guts. The pit was not a pretty sight, like something out of a Bosch painting with the smell on full.

Shortly after, a chortling Michelle came over to me with some fresh fruit: pineapple and papaya. They tasted delicious, much better than any you'd get in a supermarket. Interestingly, Michelle told me that the villages consider fruit to be a poor man's food and would rarely serve it to guests; they prefer to serve guests manioc with all the hassle it takes to prepare. That explains manioc's high reputation, whereas fruit, which grows in abundance and requires no effort, is looked down upon. I could relate to this: the more effort something is, the more worthwhile it becomes.

Michelle explained that today was the day for negotiations: "We must go to the center of the village. Those negotiations are very important. They decide the cost of the trip."

The Chief escorted us to one side of the village and bade the three of us sit next to him. All the men of the village crowded round, many of them holding their hunting spears. Then a young man appeared and began to dance, shouting as he did so, invoking the spirits to assist us. As he came towards us, he thrust the spear in our face in mock gesture. It was good-natured but impressive. Michelle said: "Put 1,000 CFA on the floor." We did so and the young man picked it up with his spear.

We then moved on, further down the village. "There are three parts to the village," explained Michelle. "This is to ask their permission to proceed with negotiations."

At the end of this spectacle, we were told to proceed to the seat of the elders, or notables. Here, under a wooden structure and carrying a spear, sat the chief elder, smeared with red paint. Despite his age, he was still physically impressive. His eyes, smoldering with passion, looked intently at us as the same young man began negotiations. With the help of Michelle, we eventually agreed a price of 95,000 CFA for the trip.

We were then told to wait at the sacred temple that was behind an old mahogany tree. A man with a horn came out and tooted our entrance. And we waited, and waited, and waited. Eventually we were brought back to the temple. Michelle turned to me and said: "Adam, the Elders are pleased with you, and they wish you to become an Ambassador for Boha."

"Tell them I'm honored, Michelle."

"Right," said Michelle, a little sniffily I thought.

Nick handed over the cash and the negotiations were done. The whole process had taken over five hours, but it had been a fantastic spectacle. I thought we would be able to leave the next day.

That night was very much like the previous one. A chicken was killed as a special treat and we again drunk copious amounts of whisky, which by now we had gotten used to.

The next day, as we prepared to leave and began unpacking the tents, the Chief came over to inform us that we would not be going that day, as they had miscalculated the number of porters they would need. Frustrated, we sat about for most of the day, doing little. In reality, as I would find out later from Michelle, the argument about porters had to do with us. Getting one of the porter's jobs was a prestigious position and there had been a lot of debate about who should go. This debate dragged on for another day and we started to get quite annoyed.

Finally, we brought the Chief over and Michelle said to him: "For these people from the West, time is money, *Monsieur Le Chief.* They have little time."

The Chief eventually agreed to go the following day, no doubt aided by the fact that we'd threatened to set off ourselves without them. Michelle had a real snob value in his attitude towards the Boha villagers. He was an extremely sensitive man, and had a huge diva-like strop. "These people are like animals!" he said and stomped off. The Chief looked in his direction sternly. It was clear from the look on his face that the feeling was mutual.

I spent the rest of the day reading and examining the small wounds I'd incurred so far, preparing them for the journey ahead. The toe of my right foot was sore. I also had a blister on my heel and an increasingly nasty looking mosquito bite. I treated it with my trusty surgical spirit. I was not as unlucky as Nick though. While swimming, he had been bitten by a type of horsefly.

"Oh," said John "that one will lay its eggs and in five years time, you will feel the larvae appear behind your eyes."

"Oh, fantastic," said Nick.

The next day, I heard a woman talking to her son as we were preparing to leave. "Can I go to the jungle, *maman?*"

"No, the jungle is very dangerous," she said as she soothed the child.

Christian, who spoke excellent English, came to see us off. I liked him but, unfortunately, he had not been selected to be one of our porters.

Our team assembled. There was Michelle and Gilbert, our

soldier; Sam, the young man on the horn during our negotiations with the elders; Sylvestre, who was a tall young notable; Marcellin, who was to act as a porter; and the Chief. As we set off, Gilbert fired a couple of rounds from his Kalashnikov into the air.

Then we entered the jungle. I found the going extremely hard as it rained on the first two days of the trek. Most of the pathways were flooded and we spent a huge amount of time wading through water, which John informed us was full of crocodiles. I was soon to learn that the plants, which seemed offer the hope of a handhold, were in fact covered in thorns and we all fell on our arses frequently.

At one point, a large green snake slid across my path.

"Green Mamba," said Michelle. "Very dangerous."

We made little progress on the first day, covering only eight of the planned fifty-six kilometers. We had also had a long stop to give the guys time to weave baskets to help them transport the plantains and manioc we had brought for the trip.

By this time, I bitterly regretted being a smoker. However, I think the low point of the journey for me were the bees. Sweat bees. Every time we stopped for a rest (and remember it was eighty degrees and similarly humid), we were covered with them. Sweating more than most, I was the not-so-proud owner of a virulent swarm. Obviously after salt, they crawled all over me and when trapped in clothing they stung all too often. Indeed, at one point, my right hand became badly swollen and I was unable to grip anything properly for two days.

Eventually we stopped for the night and pitched our tents. The guys had lit a fire and we joined them. Sylvestre, who was extremely chatty and had a good sense of humor, commented on our tents. "English sleep there. Africans sleep here," he said pointing to his structure. It was a slightly raised log platform, covered by a thatched roof. It made me feel a little guilty. I had wanted to muck in with the locals, but in the evening our sleeping arrangements separated us.

Traveling through swamp jungle is like a perennial journey through twilight. The sky is a welcome release. Underfoot though, we began to make greater progress. You change from dry kit at night to your wet kit that you wore during the day. I must confess that the only pleasure I got from putting on my wet kit was in drowning the diehard sweat bees that had refused to bugger off as I headed into the water.

We had to make thirty kilometers that day if we were to spend a reasonable time at Tele, so we really cracked on. There was no time to purify water; we drank what we found in the pools. In the trees were numerous monkeys, and although we saw signs of elephants and gorillas, we rarely saw other wildlife.

Then, suddenly, Roland, a member of our team and a hunter, stopped and pointed. A big toothy grin crossed Marcellin's face. As I looked up to see what he was pointing at, there was a large black and white monkey wanking in the tree. Obviously past the point of no return, it had not fled when it had seen us. Roland seized the opportunity, armed his rifle, and fired.

Crack! The monkey fell to the floor. Marcellin ran over to grab it.

"Look, look!" he said excitedly. It was covered in sperm. "Tonight, we eat well!"

"It won't add to the taste, Marcellin," I observed sourly. I had reservations about eating the monkey. Not because of its special seasoning (the fur, etc., would burn off in the fire before it was skinned), but because I love the creatures so much.

We shared our food with everyone, so monkey and mash it was. Apparently, the head is particularly tasty, and Sam, another member of the team who was the eldest and most revered notable, got the honors. I tucked into the ribs. It was my office Xmas party back home that night and while they were having turkey, I was sat in a swamp tucking into spunky monkey!

"Who's got its cock, Marcellin?" I asked. Everybody roared with laughter. Marcellin spent the rest of the journey teaching me African swear words.

Eventually we reached the edge of what seemed like impenetrable swamp jungle. Marcellin stopped where I had collapsed and said: "We rest here, Adam, because the last bit of the journey is very, very, difficult."

That evening we amused ourselves by using our head-mounted flashlights to locate catfish – a tasty snack. I was completely exhausted and had difficulty swallowing my food. Anyway, after eating, I went down the trail just a little and changed from my wet kit to my dry kit. I then settled down to hear the rain on the tent roof and soon fell asleep.

The next morning I awoke and, psyching myself up for the trip, downed a Mars bar, had a brew, and prepared to change into my wet kit. But where were my boxer shorts? I went down the trail to look for them. When I found my discarded boxers, I noticed that termites had been busy building a nest on top of them. I decided I didn't need them after all. I reflected on the termite's proud Queen sitting on her new throne, a mighty termite empire being built out of my pants. "You'll have the biggest tower in the district fuelled by those," I said. It was time to go. I was talking to termites.

Michelle was no master of understatement. The journey was very difficult and largely involved balancing on large tree roots, which is difficult with a pack, and moving from branch to branch. Although it totaled only three kilometers, it was the most difficult trek I've ever done.

Marcellin and Sylvestre kept our spirits up. "*Fatigué? Non!*" they would shout. We had to shout back "*Non!*" plus various other songs, which I can't recall.

With one kilometer to go, I smashed my foot against a branch and had to stop. After I rested briefly, Marcellin carried my pack for the last stretch, which helped me enormously.

Suddenly, we could see right through the darkness. "The lake is here," said Sylvestre excitedly. "We are here! We are here! Come on, Adam, come on, Adam."

When we reached the lake, Sam gestured to us all to jump in. It was part of the ritual, for which he was responsible, as a leading notable. "Quick in now. Now!" he shouted.

We all obeyed, and for the first time, I felt the warm waters of Lake Tele lap over my body. It felt weird. Standing there looking across the lake after that arduous journey, it felt like a baptism.

Lake Tele is an immensely beautiful place. Stretching some six kilometers, it forms an almost perfect circle. Jungle squeezes it all around, yet that morning it seemed to shine like silver glistening in the sunlight.

During the afternoon, we set up camp. I had bruises all over my body and spent several hours digging out thorns from my hands. Both Nick and I had blisters on our feet, probably due to our jungle boots. Jungle boots are a tight fitting boot, especially around the heel, and although they let out water (through small metal ringed holes in the sides), your feet are pretty much wet all the time in swamp jungle, so it makes no difference. Both of us resolved never to wear the bloody things again. John, who wore an old pair of squash shoes, went completely unscathed. Still, at Tele we had plenty of time to heal our wounds.

We soon realized that there would be no going on to Lake Makele though. With only five days here, we would have to concentrate on Tele. Also, Sylvestre informed me that the notables would have to give us permission to go to Makele, and they had not done so.

I had theorized that if the *mokele-mbembe* did exist, we were most likely to find it in the channels leading into Tele. John was skeptical about the existence of large channels, but we did know that although the lake was itself quite shallow, a large deep channel ran through it, as described by Roy Mackal. The water had to come from somewhere.

Meanwhile, the Chief and the rest of the guys were preparing to do some serious fishing. They had brought nets with them. This was no doubt a serious opportunity for them, as no one ever fished this lake – apart from the crocodiles. Fish were in abundance and Marcellin and Sam soon prepared a small smoke house. I was conscious that we might be eating some unknown species, but I have to confess not to have been to concerned once I had tasted it. The fish was absolutely delicious!

The next day we began our search, using *pirogues*. We were to head for the old hunter's camp where Agnana had told me he had spotted the creature. As we punted on the lake, I observed and took videos. The others took turns *piroguing*, but I just sat there, observing intently. I didn't want to risk missing anything. If I'd come all that way, I wasn't going to miss my chance. From my experience of tracking wildlife, I knew that I was likely to have just a few seconds to react, if I was lucky.

When we arrived at the old hunter's camp, the hunch had proved right: there were large tributaries running through the lake. Fantastic! Going down them slowly in the *pirogue* was like stepping back in time. We ducked under thick branches, the jungle's greenery pressing against our skin. Sylvestre deftly moved the larger branches away with his oar. Eventually, we came to a stop. We would wait and observe for a while and then move on.

One of the biggest delights of stepping onto dry land was the butterflies. Unlike the previous unpleasant company, the sweat bees, the butterflies were a real treat. Different types covered you the moment you sat down. It was like something out of a fairytale. Beautiful they were too, with their vibrant golds and blues.

On the way back to camp, Sylvestre and Sam sang songs, some lyrical and mournful, others more risqué, which caused a good laugh.

"I'll teach you one," I exclaimed.

"Okay!" they both shouted. Eventually they learned what we practiced. This is how it goes:

"Oh Manchester is wonderful,
oh Manchester is wonderful,
it's full of tits and fanny and United,
oh Manchester is wonderful!"

"Bloody Manchester United fans get everywhere!" someone in the other boat grumbled.

"Yep," I said. The Chief loved it.

Back at camp I asked Sylvestre about the tributaries. It turned out that the rivers do link all the way back to Likoula, although they are not navigable by *pirogue*. I was certain that the creature would have to enter the lake through these tributaries. As these were also confined spaces, they would give us a much better chance of spotting the elusive creature. It also turned out that each river "belonged" to a notable family. Both Sam and Sylvestre, being notables, had their own family river.

The following day, I asked the guys to take me to the old Pygmy camp. Mackal had described this in his book. The Pygmies had apparently trapped one in 1956, but everyone who had eaten its flesh had died. This was also the place where the Japanese expedition for the creature had set up their camp. John again professed that he didn't believe in the *mokele-mbembe*. It was the only time I saw Sam and Sylvestre angry.

"It is a real animal!" shouted Sam in fury. He was clearly affronted by the implied assertion that he was mistaken. Sam then turned to me and said: "But, Adam, I have to say, you are unlikely to see it, you know. It normally comes here in July and August. Most of the time it lives at Makele."

"But there is a chance," Sylvestre said. "We will do our best."

And so they did. The remaining few days were much like the others, stopping at vantage points, *piroguing*, and then moving off. The only time they became animated was when tsetse flies landed on us, or them. They would instantly sweep them off.

On the fourth day however, Sam spotted an insect flying past us that he liked. A bee. Within a few seconds we had turned our backs on the water and headed for shore. Five minutes later we were all gorging ourselves on fresh honey.

Ten minutes later we all felt like being sick. Never in all my travels have I met a man as superbly adapted to his environment as Sam. He loved the jungle and the lake. And after a few days there, I loved Tele too.

Nobody lives at Tele now. The Pygmies left in 1956, and anybody speaking to them for first hand information about the *mokele-mbembe* now would be disappointed, unless you were looking for information about sightings in the Likoula region and not Tele itself. The tribe that Sylvestre and Sam belonged to had to move to Boha after the flood. These occasional visits back to their ancestral home meant a lot to them.

With only a couple of days left on our expedition, John was getting frustrated about not being able to his research, especially as there were so many crocodiles around. Sam was also proficient at catching them. Indeed, he taught us how to make a cawing sound with our hands, swirl the paddle in the water to mimic its tail, and then lunge with the spear if one swam towards us. He caught two, which we all ate. John decided to pay Sam 10,000 CFA to go out at night when Michelle was asleep. We all discussed it, but were worried about what would happen if Michelle found out and decided to tell Agnana.

"If he locks us up, John, then I'm having your rice ration," I said.

"Agreed," said John, and off he went. But unfortunately, he found nothing.

The next morning I awoke to considerable excitement in the camp. "There, there," shouted Michelle. In the distance I could see what appeared to be a large humped creature moving across the lake.

"Quick, quick," I shouted and jumped in a *pirogue* with Sylvestre rowing furiously. The adrenaline pumped through my system. Could this be it? As we got closer, it soon became apparent that it wasn't.

A clump of weed debris was floating gently across the lake. "No, it's just what it eats," said Sylvestre, disappointed. "I have never seen it, unlike Sam. I must see it!"

"You will," said Sam, "you will. It has a horn, Adam, he will see."

Back at camp, there was some consternation. Michelle was highly suspicious of John and Sam and remained surly for the rest of the day. We were worried what he would tell Agnagna. Eventually though, he did respond to some conversation, and I assured him that John had no opportunity to do anything.

It was our last day at Tele. We'd seen plenty of crocodiles, birds, snakes, and butterflies, but no *mokele-mbembe*. Even locations suggested by a dowser named Hans Operdorff had brought no joy; yes, I'd try anything. Now it was time to go back. Five days was not enough time to complete a search, but if we were to rendezvous with Mr. Paco and our boat, then we needed to go.

As a parting shot a huge silverback gorilla had climbed to the top of a tree to look at us. He sat there picking his nose nonchalantly, ready to see us off. He was a magnificent creature worthy of his surrounds.

So we loaded up our stuff and made for home, which was the village.

"Adam, I miss my wife, it is time to go," said Marcellin.

"Don't worry, you'll see her soon," I said.

Sam then chipped in: "I miss my wives."

"How many have you got?" I asked.

"Three, and sixteen kids," he replied. "And Sylvestre, he will soon marry my daughter." Sylvestre didn't respond and looked at me blankly.

The journey back to the village was a dream compared to the

journey there. Thanks to the dry weather, many of the paths that had been flooded before were now completely dry. As a consequence, we made good progress. The villagers, as usual, walked barefoot, while I squelched along in my jungle boots.

On the first night, Nick was knackered and we stopped to eat a small diker (deer) that Roland had caught. By nightfall on the second day, we had made it back to the village. As we arrived, we were given a huge reception; all the villagers had turned out to meet us and cheer our arrival. As Sylvestre walked proudly beside me, I asked why we were getting such a glorious reception.

"Oh, because the villagers consider you're their *bon ami*, Adam," said Sylvestre. "You have earned their respect."

"But why did they show little interest when we first arrived?" I asked.

"Oh, we knew you were coming," said Sylvestre.

"Knew we were coming? How?" I said, surprised.

"Oh, my father saw you were coming in a dream," he answered, quite matter of factly.

"I must spend sometime talking to your father," I said.

"Tomorrow you shall be a guest at our house," he said. With that, I climbed into my tent and, exhausted as usual, fell asleep.

The next morning I awoke and began my normal routine: examining the cuts I'd incurred on the journey back. There were a few small ones, but my septic scar had healed nicely. I was hobbling around, however, as I'd twisted my foot on the way back. The toe I'd smacked in the swamp was now black underneath the nail and the nail was clearly due to drop off. Otherwise, I was in one piece.

During my time in the jungle I had not bothered with purification tablets. We'd had to move quickly through hard conditions, and the water had been fairly acidic (there are no leeches in the Congo swamps). Besides, I'd become sick of drinking purified water. It's rather like downing the contents of a swimming pool. (There were no neutralizing tablets available at the time of this expedition.)

I spent some time mooching about the village, reflecting on my adventures. Eventually Sylvestre came to get me, and we headed off to his father's house.

"Adam," said Sylvestre, "try and say I have seen the *mokele-mbembe*."

"Why?"

"It will earn me great respect," he said.

"I will say that you were a great help to me at Tele, and an honor to your family," I replied.

"Thank you, Adam."

At the house the old man greeted me, shaking my hand furiously.

"And how was your journey to the lake?"

"It was great, really magnificent. But I did not see the *mokele-mbembe*."

"I know, I speak to it," he said. "It is now at Lake Makele."

"You speak to it?"

"Yes, I speak to it in my dreams," he said.

"But have you seen it?" I asked.

"Yes, I have seen it many times."

"What does it look like?"

"It has feet like an elephant's and a neck like a giraffe. The male has a horn; the female is without. It does not live in the water; it travels through it. It lives in the jungle. You will see it one day, Adam. For you are strong with the spirit. And I, I understand the spirits." Then he smiled.

In the corner of his house, I noticed what looked like a giant wheel with strange symbols behind it, partially obscured by a curtain. "What is that?" I asked him.

"It is for the spirits, Adam. The *mokele-mbembe* is a sacred animal, which is why the Pygmies died when they ate its flesh. That is forbidden."

During my conversation with the old man, I learned about the society in Boha. Apparently there are clear divisions. The chief and the secretary function as administrative officials; they are elected every five years. The notables, who are almost like an aristocracy, are born into their roles; they function as the mystical and spiritual base of the community. The two sides are not interchangeable. A notable cannot become a chief or secretary or vice versa.

"If you lived here, Adam, you would be a notable," he said.

"Thank you, sir. I am honored."

"There is no need. You are what you are."

After we bade our goodbyes to Sylvestre's father, Sylvestre took me up to the school. The head teacher came out to greet me and all the kids poured out after him.

"*Bonjour, monsieur,*" they shouted in chorus.

"*Bonjour, mes enfants,*" I replied. Sylvestre and I played with them for a while, chasing them around, which they all enjoyed. He also indicated to me that some of the money we had given for our expedition would be used for schoolbooks.

That night, we settled down for a delicious goat stew, following by some dancing, but it was more muted and melancholy than it had been earlier. I asked Sylvestre why this was.

"It is because a young woman has died of the sleeping sickness," he said sadly. "She was only twenty-six. Many, many people die that way, more than from any other thing." *The tsetse fly*, I thought to myself. It was the only thing that made them all flap their arms in the jungle, and that was why. The village had a wonderful community spirit, but this occurrence served to remind me how lives could be shortened by virulent disease.

The following day, our penultimate one in the village, was a cause for celebration. A large, old, Mississippi-style steamboat named *The Almina* was chugging along in the Likoula near to the village. Amid the great excitement, we joined the crowds in the *pirogues* and rowed out to greet it. Clambering aboard, I found that their supplies were low. To my disappointment, they'd run out of cigarettes. However, they did have some bottles of German beer, which would be a welcome release from the village whisky.

The next morning I went to see Michelle. We had a good chat and talked about the journey back. Gilbert also joined us. "Manioc, manioc, manioc. I hate manioc, Adam!" he said. "I have had no bread for weeks! Once we get back to Impfondo I shall take you to a restaurant, we shall have good food and then go to a club and drink beer."

"Great," I said. "That sounds fantastic."

The boat was now late. Standing by the shore, I saw a craftsman fashioning a new mahogany boat. *God, those trees take so long to grow*, I thought. But to lecture these people would be greeted as patronizing. They needed a reason to preserve their environment.

I had enjoyed my time in Boha, but it was now time to leave. Eventually the boat came. It was not Mr. Paco this time but

Michelle's brother. All the villagers turned out to see us off. Marcellin gave Nick a gourd he had seen in his house. Sylvestre and his father gave me some sugar cane and bananas.

As the boat was pushing off, the old man said to me: "Adam, you must return."

"Forgive me, but I cannot promise I will be back."

"You will return, and you will go to Lake Makele next time." He gave a deep throaty laugh, and then he was gone.

The journey back was pleasant enough, first stopping at a conservation project that encouraged the locals to plant sustainable crops in the environment. Later we went to the village of Etangi and had a drink with Michelle's sister. Finally, we arrived at Epena.

The house where we had stayed a few weeks before was occupied. It had been cleaned up, painted, and furnished. The "garden" had also dried up, as the waters had receded in the dry weather.

We began unpacking our gear on the ground outside the house and started making a brew with some hexamine we had. Suddenly the door of the house swung wide open and a bearded, middle-aged man said in a thick American accent: "When you Brits have finished squatting in the dirt there, why don't you come in and have a beer?"

"Cheers," we all said. The American turned out to be Brian Curran, who headed World Conservation International in that part of Africa. They were there to help John and Connie organize the Lake Tele reserve. So, Tele was to be tamed! I settled down in an armchair and had a coke rather than a beer, although he had both. I swear, I would've paid 50 pounds for a cold one that day. They were genuine friendly people and I admired the work they were going to do at Tele, though I could foresee problems with the villagers of Boha who regarded the lake as their property. I hoped the experience wouldn't turn sour on them.

The next morning as we awoke, Nick pulled a millipede from his tent.

"Fuck!" he shouted.

"That's the only thing that's been in your trousers all trip," I exclaimed.

"Very funny," he said.

My cuts and bruises were healing well now. One aspect of living in a village that was less than wholesome was the damn flies. These flies, which normally pester cattle, specialize in attacking wounds. They are equipped with small barbs that puncture wounds and at the same time secrete an anticoagulant to stop it from healing. Thankfully, these unmitigated bastards were not present in Epena.

Brian Curran gave us a lift to the missionaries, as they were going to stay with them as well. Once there, I gave one of their children, Hap, a kickboxing lesson, while Nick went to check on our reservations for the trip back to Brazza. We didn't want a repetition of our journey up to Impfondo, especially as by now it was nearly Christmas and we were keen to go home.

That night we all trundled off to a restaurant, passing as we did so many trees filled with fruit bats that chatted noisily to one another in the moonlight.

"*Ici*," said Gilbert excitedly. The restaurant was a small wooden shack. At the back of it, Gilbert had arranged for a table for us, which was covered with a plastic sheet. We sat down and waited. "Look, Adam, we have *du pain!*"

"The bread," we cheered. "No more manioc! Hooray!"

"What are we having with it, Gilbert?" I asked.

"It is the bread *avec du singe!*" he exclaimed.

"Bread and monkey?" I asked, disappointment swelling in me. "Is that it?"

Gilbert looked at me quizzically. "Yes, c'mon, Adam, meat from the monkey is very good." Gilbert had selected the ribs for me as a special treat. I had to make the effort, otherwise he would have been offended.

On the table I spied some Tabasco sauce. As I began pouring it on my meal, Gilbert lunged across the table to put his hand over my bowl. "No, no, no. You will ruin it, Adam!" he said earnestly.

"Yes, Gilbert," I sighed. "You are right."

After dinner, we headed off to the club. I doubt I've ever been keener for a beer. I needed to wash down my Tabasco monkey sandwich. The club was busy and we all joined in the dancing. Gilbert kept to his word and introduced me to a young lady. I bought her a drink but as politely as possible left her company at the first given opportunity.

"Do you not want to boom-boom, Adam?" said Gilbert, laughing.

"Not tonight, Gilbert, I've got a headache" I replied.

"You are strange, Adam."

"Yes, to you, Gilbert, no doubt I am."

In the middle of the night, I was awoken by the sound of gunfire. The missionaries were increasingly worried by the violence on their doorstep, especially with the kids there. I don't blame them. Their determination reminded me of how little we sacrifice for others. After having fried eggs for breakfast, we offered to work in Joe's hospital for the day. This was to be one of the most sanguine experiences of my life.

After a tour of the little hospital, Joe had us counting out prescription doses, while he dealt with his never-ending queue of patients. Joe works under pressure. He did once have a partner, but as I understand it, the guy died of malaria at the age of 36.

In the afternoon, a young woman in terrible distress walked in with her little baby.

"The baby has AIDS," said Joe. "And she will die of it too." As the mother lay on the table, she unwrapped her clothing to reveal a huge abscess twice the size of her breast. She was crying and in obvious distress. A man whose face was distorted with leprosy helped to hold her down, while Joe cut open the abscess and drained a green fluid from it. I tried my best to comfort the woman, stroking her brow and popping a sweet in her mouth. She was extremely brave, and I felt deeply moved by the scene. Later, Joe told me that he needs 250,000 dollars for his hospital. I hope he gets it. Fifteen percent of the population of Impfondo has AIDS, but there are other less well-publicized diseases, such as sleeping sickness and leprosy, that afflict them as well.

The "games night" at the missionary station was a welcome relief. Afterwards, we sang "Silent Night" around the Christmas tree. It felt strange singing Christmas carols in Impfondo and my mind turned towards thoughts of home. Our plane home was due to arrive the next day. If we missed it, we wouldn't be home for Christmas. But if worst came to worst, we'd arranged to go Bangui in the Central African Republic and try to catch the place from there.

The plane did arrive the next morning. At the airport we said our goodbyes to the missionaries and Gilbert and Michelle also turned up to say their goodbyes to us. As I boarded the plane, Joe came over and said: "I wonder if you could do me a little favor, Adam?"

"No problem, Joe, what is it?"

"Could you post this for me, this package?"

"Sure, what is it?"

"Some tapes. The post here is, as you might expect, unreliable and our folks would like to hear our kids sing."

"No problem, mate, I'll post it as soon as I get home."

As I left I gave Gilbert something: the little England bag with my son's name on it that I had used to carry my food. Somewhere now a Congolese soldier will be trekking through the jungle with that bag.

The next two days in Brazzaville passed fairly slowly and I kicked my heels. I made a visit to one of the most expensive hotels in Brazza where rooms are 250 dollars a night. It seemed to be occupied almost entirely by French and Lebanese businessmen but it had an excellent swimming pool.

By now my guts had started to feel rough, a situation made worse by the fact that I was delayed one night by problems with the Air Afrique flight.

Eventually I arrived back home in Manchester and celebrated Christmas with my family, which was a great relief to my mother. As we gathered round on Christmas Day, I asked them all to sing "Silent Night." I thought of the missionaries back in the Congo and, of course, of the old notable who'd said I'd return.

I learned that I had contracted Gardia, a tropical parasite, and scabies to boot. But a week later I was fine. Right after Christmas I began planning my next expedition. I didn't know it at the time, but the results would be sensational.

Chapter Eight

Orang-Pendek

The *orang-pendek*, which means "short person" in Indonesian, is said to live in the dense mountains jungles of Sumatra. Stories about the Orang-Pendek go back as far as Marco Polo, who wrote about them in his diaries. Even then, these "little people" were regarded as elusive. The Orang Kubu, the forest people, have regarded the *orang-pendek* as a real people for centuries. This ground-dwelling, bipedal primate is normally described as being short, no more than five feet tall. Its fur varies in color from a dark orange to a chocolate brown. Its "human-like" face is said to startle, and even upset, witnesses, such is the intensity of the experience.

Dutch colonists also recounted seeing this creature. One of the most frequently reported is that of a Mr. Van Heerwarden, who described an encounter he had while surveying land in 1923: "I discovered a dark and hairy creature on a branch. The Sedapa was hairy on the front of its body, the colour here was a little lighter than on the back. The very dark hair on its head fell to just below the shoulder blades or even almost to the waist. Had it been standing, its arms would have reached to a little above its knees; they were therefore long, but its legs seemed to be rather short. I did not see its feet, but I did see some toes which were shaped in a very normal manner. There was nothing repulsive or ugly about its face, nor was it at all apelike."

Before I began planning my expedition for the *orang-pendek*, I reflected on what I had learned from my previous expeditions, as I always do. First, it was important to have a common objective. Having two separate goals, as in the case of the Congo, hadn't worked. All those on the expedition must have the same motives, though sidelines are okay.

Second, it's always good to have at least a core team of individuals around you, who, if all else fails, you know you can trust. That way you know one another's strengths and weaknesses.

Lastly, all team members should have an optimistic disposition. Personally, I realized that I needed to improve my camp craft and fitness skills, both of which I worked hard on from Christmas until we left the following September. I gave up smoking and did more cardiac work than I'd ever done in my life.

Our Congo expedition had generated worldwide publicity. This may have led to false expectations among our hosts about our financial clout. For Sumatra, we would be fairly low key, not least because the Indonesians are very sensitive about how their country is portrayed. Besides, the expedition has always been the primary motive for me, not the publicity. Again, I was doing the trip on my own money. Cash was really very tight, but barring safety, I didn't want anything else to stand in the way of a good adventure.

Both Keith and Andy had expressed an interest in going to Sumatra, and I knew I could trust them. Thus, we decided we'd keep the British team small. There was no recruiting on the internet – that's just too risky.

I had the advantage of having been to Sumatra before, so I understood a little of the region and its peoples. Sumatra is extremely tribal, each area having its own unique, proudly defended culture within a relatively short distance of one another. A good example is the Minangkabauare, the world's only Muslim matriarchal society. Within their community, property is passed solely through the female line. That means, if my parents died, my sister would automatically inherit the property, and I would receive nothing. Contrast this situation with the predominantly Christian Bataks, where males rule the roost. The Christian Bataks and the Minangkabauare live literally back-to-back on the island, yet have very different traditions. But there is no evidence of any animosity between them.

I had read very little about the *orang-pendek* before my first visit to Sumatra. I was traveling through the country, when at Kerinchi Seblat National Park, I decided to investigate the story of the *orang-pendek* and took a guide, whose name was Sollok, who had claimed to have seen one. His story really intrigued me. He had been on a path when he spotted two, small orange black primates, as he described them, who were fighting. When he approached, they took one look at him and bolted off into the jungle. He pursued them but lost them as they ran over a ridge and disappeared into the jungle.

Sollok knew the creatures in the area intimately, and he is one of Kerinchi's most experienced trackers. He could not possibly have mistaken them for something else. After all, Sollok was less than three feet away from them! With first hand accounts like this, you either believe or you don't; he's either making the whole thing up or he's telling the truth. Forget having made a mistake; mistakes are unlikely and are used as an excuse to explain away sightings like this by cynics.

Sollok took me to the spot where he had seen the creatures. As we trekked through the jungle, he gamely sprinted over a log with a two hundred foot drop on the other side. He was extremely fit – yet these creatures had managed to outrun him. Having heard his story, I decided to contact Debbie Martyr when I got home. She is the British journalist who had made it her business to prove the existence of the *orang-pendek*.

Debbie has been associated with the *orang-pendek* more than any other person in the world. She claims to have seen the creature twice. As a result, she had launched Project Orang-Pendek together with the photographer Jeremy Holden. This mainly consisted of setting camera traps in the jungle, hoping the creature would walk through them, and thus be photographed. The project had run for five years with only limited success, unfortunately. No definitive proof had been found. This had led some people to question their existence. Were they real or flights of fancy? Was she making the whole thing up, or was there something to it?

Nonetheless I was willing to put my money on Debbie. I thought the *orang-pendek* might be a winner. But even if it did exist, it was most likely an extremely rare and elusive creature, so it would still be like looking for a needle in a haystack. There was a chance though – and a chance is always enough for me.

So we began negotiating with Debbie and arranged to hire some of her tiger team. She was now running a project aimed at stopping tiger poachers in the area. Sumatran tigers are beautiful but rare creatures under intense threat from two sides. Not only did they have to fear poachers looking to make a fast buck, but their environment was quickly eroding. As in the Congo, big money can be made from poaching, and social pressures can encourage people to kill animals that even they revere. It's not enough for Westerners to simply say "no" to these people. They'll just laugh at you when you're gone. You need to provide them with incentives to protect their environment, which is often their most precious asset. Debbie was working on this as well, so I had great respect for her.

Debbie knew a man named Carlos, who had a minibus that could take us up to Kerinchi. After arriving at Sungai Penua, we would then trek to the area of Sollok's *orang-pendek* sighting and make camp, spending roughly two weeks in the jungle. Remembering my return from the Congo, we also arranged to spend a few days in Singapore at the end of the expedition to chill out.

Both Andy and I were pretty fit at this point, and Keith had started his fitness program. Keith and I went for many long hikes in the Derbyshire hills not far from where we live and planned our trekking methods. Keith increased his load each time he went, eventually completing several hikes in full kit.

The plan was for Keith and I to fly from Manchester, Andy from Heathrow, and we would all rendezvous in Singapore. Before we left, we also decided that I would lead the expedition, Keith would be in charge of expedition finances, and Andy would deal with the logistics, as he had been heavily involved in liaising with Debbie before we left for Sumatra.

So Keith and I left on the morning of September 10th and arrived in Singapore 17 hours later. We walked over to Andy's flight straightaway, which was due in ten minutes after ours. Although we were knackered after our flight, I wanted to waste no time in getting the money divided up. Before we had left, Keith and I had worked out roughly what we would need for our expedition. I wanted to exchange our money into Rupiahs, U.S. dollars, etc., right away, as I didn't want to be seen handling large amounts of cash as soon as we hit Indonesia.

Keith had just finished noting everything in his accounts book and stashing the money for Carols and the minibus in his pocket to keep it handy, when Andy's phone started clicking out a text message. Andy read it and looked stunned.

"What's the matter," I asked?

"I've just got a message that part of New York's been blown up, the Twin Towers where I was last year."

"Unbelievable!" We had landed in Singapore on September 11, 2001.

We had about five or six hours to wait for our flight to Pedang in Sumatra, so we rushed over to the TV screens where they were playing CNN. As we watched the story unfold, a bloke with a South African accent standing by my side turned to me and said: "There's going to be all sorts of shit to pay for this."

I didn't reply. Like everyone, we all felt a deep sense of shock over the events of that day and chewed it over for many hours. All air flights were starting to be cancelled and we realized that we might face a completely different world when we returned from the jungle.

To get our minds off the nightmare, we booked a short city tour of Singapore. It would also give us a small taste of the city that we were to see at the end of the expedition. Singapore is an impressive city. Due to the lack of space, the skyline is dominated by pristine skyscrapers. Virtually everyone lives in a flat, but they're not moldy high rises like some of the inner-city constructions in the U.K. There was no graffiti; indeed the whole place looked like some metaphysical cleaner had just scrubbed it. The Quays area, with its many bars and restaurants, looked highly cosmopolitan.

Returning to the airport, we boarded our short flight to Pedang and arrived without delay. We unloaded our luggage and stepped out into the baking tropical heat.

We had arranged for Carlos to hold a sign up with our names on it, but surprise, surprise, there was no sign anywhere to be seen.

"Shit," said Andy. "What do we do now?"

"We'll give these guys half-an-hour and then negotiate a price with one of the drivers here," I said. "We have to get up to Kerinchi tonight, as we start paying for our guides tomorrow." As we were discussing this though, two small men came puffing towards us.

"Adam, Andy?" said the smaller, spiky-haired guy.

"Yes."

"I am Nicky and this is Amir," he said, introducing a slightly taller, chubbier individual. "We work for Carlos. Sorry we are late, we got the times wrong."

"No problem," I said, "but we are very tired. We need to get up to Sungai Penuh now."

"Ah, yes" said Nicky, "but Sungai Penua is very far, through the mountains, seven hours or more. You must rest first. We go to Carlos's and stay the night."

"I don't want to stay the night, Nicky." By now I could see which way this was going. "I want to go now."

"How much you pay?" said Nicky.

"Three hundred thousand Rupiahs, as we arranged."

"That is too little. It is six hundred thousand each way and we go tomorrow after you stay at Carlos's."

"Where is Carlos? I think I should speak with him," I said, raising my voice. Nicky looked surprised.

"We will go now, Adam. Do not worry."

I spoke to Nicky on the journey down. It seemed that Carlos was a big man in the area with financial fingers in many pies. But there was no shortage of competition for drivers here. I wasn't a wetback and he wasn't going to screw me.

After an hour or so, mostly through heavy Pedang traffic, we arrived at Carlos's. By now it was night. I have to say, Carlos's beach hotel was extremely attractive. Palm trees and a lovely beach fronted it, and there was a bar and a rather pricey restaurant. Looking out to the bay, there were small fishing boats adorned with lanterns and lit overhead by a pale moon, in short, an artist's delight. But I was in no mood for watercolors.

We grabbed some cokes and waited for Carlos. And waited. Eventually, he appeared. He was in his forties and wiry. He fixed me with his eyes straightaway.

"You are happy, Mr. Adam?"

"No, Carlos, I want to go now, like we arranged."

"You stay here tonight; we go tomorrow. One thousand two hundred Rupiahs," spat Carlos.

"Don't fuck with me Carlos. I can get someone else."

He smiled. "Not tonight. You cannot stay anywhere else now."

"Okay," I said. "One million. I'll pay you six hundred thousand now and four hundred thousand on my return, but we have to leave now."

"Agreed," said Carlos. "You leave in an hour. Amir will go and get gas."

It was worth it because if we didn't leave then, we'd lose a day of the expedition and costs.

"Good man, Adam, You are tough, eh?" he started chuckling. "You will have no problem in the jungle? Your boys tough, too?"

"Yes," I said.

"Good, because the mountains are hard and cold, hee, hee." With this, he shuffled back into the shadows.

While waiting for Amir, we spoke to some of the backpackers who were staying at Carlos's. Several of them were preparing to leave Indonesia, worried about how the local Muslims might react to the crisis of September 11th. After all, Indonesia is the largest Muslim state in the world.

One guy, Scott, who was there with his fiancé, told us about his encounter with an orangutan. "I can't believe you guys will be in the jungle for weeks," he said. "I'd had enough after one day. I went to see the orangutans and one started chasing me. I remember thinking, why don't you just sod off and eat fruit or something? Just leave me alone!"

"Oh, where we are going there are no orangutans, just rhinos, bears, and tigers."

"You're welcome to them," said Scott.

A slight digression here, but if you ever get the opportunity, do go and see orangutans in the wild. Seeing a bright orange giant fur ball moving through the jungle above your head is one of the most bizarre experiences you can have in the wild. It's great fun, like seeing a giraffe run. Orangs are also highly intelligent; researchers have taught several of them to communicate through sign language.

Anyway, Amir duly arrived and Keith, Andy, Nicky, and I climbed aboard the minibus. As we started off, I began to relax and offered everyone a Mars bar, which they duly consumed.

"Before shut eye, Ad, a quick question," said Andy. "When we met Amir you seemed a little surprised. Why?"

"Oh, that's because you rarely see a fat Indonesian. You will see that I mean."

"Why's that?" said Keith.
"Because they don't have any burgers!" I said.
"We'll see," said Keith with a determined grin. He claimed to be able to locate a burger anywhere on the planet.

We rattled on in the minibus, but there was nothing to see, for by now it was pitch black outside.

"I have a Bob Marley tape," said Nicky.

"Great, put it on," said Keith. Thus Bob played in a continual loop for the next seven hours.

Two hours into the journey something bit me, I killed it, and fell back asleep.

Two hours later, Amir took a well-earned driving break and we stopped at a Pedang style restaurant. It was my first opportunity to introduce the guys to Indonesian food.

"What's that?" said Keith pointing to some beef rendang in the window. "It looks like mud."

"It's like a chillied beef. It keeps well, so it's useful to take into the jungle. Try it."

"No thanks," said Keith. "I'll have my pepperoni."

Now, I love Indonesia; I respect both the people and the culture, but I have to say I find their cuisine about the worst in the world. Plus, I probably got the worst case of diarrhea I've ever had the last time I was there, which no doubt tempers my judgment of their cuisine.

"You're going to get sick of pepperoni," I said. "You'll have to try the food sometime. Plus you can't waste your jungle rations needlessly."

"I'll wait, thanks," said Keith.

We finally arrived in Sungai Penuh at three in the morning. We woke the poor owner up and Nicky and Amir helped us unload our rucksacks. Keith paid them. We got into the hotel and after a short babble, settled down. We'd made it on schedule. Tomorrow morning, we'd meet Debbie.

Andy woke us the next morning. I opened the curtains of my little hotel room and looked out on Sungai Penuh. The main road was thick with traffic, small scooters purring incessantly down the main street. People were also milling about the shop fronts, several of them dressed in jumpers, although it felt hot to us. That's because Sungai Penuh is at a higher elevation than the coastal areas, and to Indonesians it really was cold. We could have wandered about in our t-shirts all night. Of course, Andy wanders about in his t-shirt in winter in England, but then he's from Newcastle.

"Are we rested then, chaps?" I said.

"Yeah, not bad," said Andy.

"I've not got the grip of this bloody mandy thing yet," said Keith. A "mandy," is the nickname for the washroom function in Indonesia. Basically, it's a large water tank from which you pour water over your head using the plastic bucket provided. You also use this to flush out the squat toilet.

"Well, you don't need worry," I said, because soon you'll be able to go *au naturelle* in the jungle."

We waited for Debbie for an hour or so, but there was no sign of her. As I said, it's always important to get a contingency in place as quickly as possible. I knew we could get guides for Kerinchi from the park office, so Andy and I decided to head up there, leaving Keith to wait for Debbie. We needed permits anyway, so the journey wouldn't be wasted.

We'd barely walked ten yards when a scooter pulled up beside us. Perched on it was a Kerinchi Park Guide, instantly recognizable from his green uniform.

"Hello, mister," he said. "You need guides?"

"Not just yet, thanks. We have guides arranged with Debbie Martyr."

"Ah, Debbie, she is a friend of mine. I am Affmir, one of the chief guides here."

After making introduction, he asked if we were going to the Park Office. Since we were, he said we could pick up permits with him, so off we trotted with Affmir, who spoke very good English, a rarity in Sungai Penuh.

Affnir told us about the trips he did for tourists in the area. "Normally they are for three or four days. We go to Kerinchi, or I can arrange for you to see the rhinos."

"We are going for two weeks," I said.

"Two weeks!" he exclaimed. "You are mad. I would not go for

two weeks." He then introduced me to another man, slightly taller and older than himself. "This is Eddie. He is a good man. He also speaks English. He can guide as well."

"Excellent to meet you, Eddie," I said. "Affnir, if we need some guides again, I'm sure we'll come and see you."

"Yes, you must, for I have my son Angry to feed!"

"Angry?" I said. "Why do you call him that?"

"Well, before I was married, I went to see my wife in her village at night. The villagers caught me and I thought they were going to kill me, Adam," he said, waving his arms around wildly. "They said I must be married that night to save her honor! I agreed. Now I call my son Angry."

"Fantastic," I said. "I should have called mine 'Beer,'" I said to Affnir.

"C'mon," said Andy, who was having difficulty talking and laughing at the same time. "We need to get back to Keith."

We arrived back at the hotel to find Keith downstairs with the hotel owner.

"Debbie's turned up. She'll be back in half-an-hour," said Keith.

"Okay, let's use the time to change some extra money. We need to make up the difference in Rupiahs from the money we paid Carlos."

But the bank hadn't had many customers in from England. At first, they politely declined to serve us, thinking we had Travelers Checks. However, after patiently explaining we had cash, the transaction was completed and we were honored: our cash duly arrived on a blue velvet cushion!

Shortly after we arrived back at the hotel, Debbie swept in through the door. Debbie is about five feet six inches. She must have been a beautiful woman in her twenties; now in her late forties, she was still quite attractive. She has long, raven-colored hair, an olive-colored skin, and is pretty thin, no doubt from eating that Indonesian food. She is wonderfully extroverted and can talk for England – or Indonesia! Her voice is quite southeastern BBC announcer. Standing between us with our shaven heads, we no doubt looked like her hired guns.

We loaded out gear into her jeep and headed back to her place, where we found all the supplies we had arranged in pretty good order: industrial quantities of rice and noodles propped up against the wall. We'd brought plenty of our own rations, of course, but not enough for two weeks. This lot was to feed eight people.

"Oh, and we've brought you a present," I said. "A couple of liters of your favorite gin." Andy and I plonked it on the table.

"Boys, I love you!" said Debbie with a grin on her face.

Over coffee, I asked Debbie about her sightings of the *orang-pendek*.

"I have seen the creature twice" she said. "Once in the area where you are going. It was literally in front of me, hiding and watching me. It's very cunning." She chuckled slightly, and took another deep drag on her cigarette. Debbie smokes like a chimney, as do most people do in Indonesia.

"How do you suppose this creature got here?" I asked her.

"Fifty thousand years ago, as you know, this island was ripped apart by volcanic activity – the northern part of the island was virtually destroyed. The southern part, where many of the tree dwelling primates were, like the orangutan was ravaged. My view is that because many of the forest canopies were destroyed, the creature had to learn to walk along the ground. I think the *orang-pendek* is a version of the orangutan. As you know, there are no orangutans in this area. One important fact, though: if you do see an *orang-pendek*, you'll be disturbed, for it walks like a man."

"What do you mean?" asked Andy.

"Well, you know how a chimp or gorilla has this particular gait?" said Debbie, who was now shuffling like a gorilla with her back hunched and her knuckles by her knees. "An *orang-pendek* walks normally. It's very disturbing the first time you see it, I can assure you."

The Tiger Team was due that afternoon, so we decided to go down to the market to buy lunch. Markets in Indonesia are always a great curiosity, rivaled only, I think, by those in Canton. Almost every variety of foodstuffs can be bought here. I took to the quail eggs; they're delicious and full of protein. The market is also an amazing place to people-watch. Where else can you see a goldfish de-scaler?

By later afternoon, the Tiger Team hadn't arrived and Debbie was growing anxious. "It's not like them. They're always here on time. I do hope something hasn't happened to them."

"What could have happened?" I asked.

"Well, they are traveling through an area where many of the tiger poachers we've had arrested come from. They could have been shot or anything."

But Debbie needn't have worried, for no sooner had she finished the statement, the team bounded in through the door. Debbie ran over to them and began jabbering in Indonesian.

"Thank God. Thank God," she said. "It turns out that it was only the fan belt on the jeep that broke."

She then began introducing us to the team: Alan (Alip), who speaks good English and leads the team; Sudaraham, our cook; Sansul, a guide who looked like a warrior and sported one of those long thin pointed moustaches; and Pendi, who was the tallest of the men and the only one who was armed.

"Oh, I just love Pendi. He's fantastic!" said Debbie. "You know not so long ago four poachers jumped him; he put two in hospital and the other two ran away. He's a bit of a reformed ex-gangster, handy in this place." Pendi gave us all a big grin. "When we get to the village at the edge of the park, we'll also be picking up Sahar and his brother John. Sahar is a great guide and John is learning off him. They also speak some English." I would soon learn just how great a guide he is.

Debbie suggested we stop off at the local pharmacist to stock up on extra medicines. "They have a particular nettle in the jungle there that is really unpleasant," she said. "Its sting, how shall I put it, is more robust than an English one and it can last for over twenty-four hours."

"Lovely," said Keith sarcastically.

After we'd bought some extra supplies, including some extra Anglo-Indonesian dictionaries, I asked Debbie about Affnir and Eddie. "Oh, Affnir," she said theatrically, rolling her eyes. "Yes, he does have a son called Angry. Actually, it's not that unusual for Indonesians to take the names of famous people. Once in Sungai Penuh, I broke up a fight between two small boys – they turned out to be called Saddam Hussein and George Bush!"

That night we stayed at Debbie's and slept on the floor. A

thunderstorm lashed at the roof all night, and I prayed we would not have two weeks of that. The next morning, we awoke to fine weather, however.

"Morning, honeys," said Andy jokingly.

"Morning!" said Debbie, obviously thinking this was directed at her.

Andy blushed and I laughed. "You're in there," I remarked.

As we clambered aboard the jeep, I sat next to Alip for the drive to Sahar's. Alip was gregarious and chatty.

"Adam, much of this area used to be jungle only a few years ago," he said, pointing at the open fields. "Now it is gone. Nothing. Very sad."

"I agree."

"We will have nothing soon if this is not stopped." Alip shook his head sadly as we rushed past the endless paddy fields.

Eventually, we arrived at Sahar's house. Many of the houses in this village were wooden, some more than one story. Sahar's was a new construction, made of breezeblock type material with a tin roof. It was only partially finished, Debbie explained, with additional bits being added when Sahar had the money. The windows were yet to be completed.

We waited in his garden and Sahar duly came out, followed by John, his brother. Sahar had thick black hair and wore small, horned-rimmed glasses. He stood about five feet three inches and was in his early thirties. John was slightly taller and more muscular than Sahar.

Using our maps, I plotted out what I proposed to do to track the creature. Basically, the plan was to move from one base camp to another in the jungle on the side of the lake where many sightings had taken place. We would spend a few days in each base camp before moving on, knowing that we would see nothing by constantly moving through the jungle. Sahar agreed to the mechanics of the plan.

It was time to divide up the weight for the climb. We stood in the shade and looked up at the jungle-covered horizon above us. Sahar's village is set in a beautiful location. On one side is Gunung Kerinchi, a gargantuan smoking volcano. We were, however, to climb Gunung Tujuh and head for the highest freshwater lake in

Southeast Asia, climbing the ridges beyond it.

"It looks a fuck of a long way up," said Keith.

"Eight thousand feet or so," said Andy. In fact, it stands 1950 meters to the lake.

"Well, there's no getting round it," said Debbie, who was standing out in the sun, saying it was too cold in the shade. I think she'd been in Indonesia too long.

Finally, we were ready to go. "Good luck," said Debbie waving us off. "Prove me right and find the damn thing!"

"We'll do our best," I said.

The path up Gunung Tujuh is steep. Normally, before you make such a climb, you need to adjust to the environment. We were jetlagged, had not adjusted to the altitude or to the tropical conditions, and were carrying full kit, weighing between 35 and 50 pounds, depending on the individual. But we wanted to maximize our time in the jungle, so we decided that we'd make the climb as quickly as possible, build our shelters, and then rest for the remaining part of the day. The rains had made the paths heavy work, however, and after a couple of kilometers, Keith looked knackered. "Ad, I'm completely fucked."

"Don't worry mate, we'll rest up, get some water on board, and relax," I said confidently. But I was worried whether he'd make it or not. Pendi led the way and I followed with Andy behind me, then Keith. I kept looking behind me so that Keith was never out of my sight.

Eventually we came to a point where the angles were over sixty-five degrees and we were dragging ourselves up with the benefit of tree roots. The gaps waiting for Keith had by now enlarged considerably. When I looked behind me to see Keith, I saw his face turn from red to white, as if he were going to pass out.

"He's not gonna make it up here, with that pack, Ad," said Andy. I agreed and we discussed our options. Finally, I decided to offer Pendi, Sansul, and Sudaraham an extra ten thousand Rupiahs each if they would divide Keith's kit up between them. They agreed and eventually we got to the top and let out a large cheer. I dished out sweets to everyone in celebration.

At that moment, a beautiful fish eagle swept down across the lake and effortlessly picked up a fish before heading east.

"A good omen," said Andy.

"Let's hope so," I said.

"Now, Mr. Adam, we'll call the 'fish people' over to us," explained Sahar. "They have canoes that can take us over to our camp."

Some locals who had also climbed the mountain had lit a small fire. We joined them and posed for photographs for them. After an hour or so of calling though, there was no sign of the fish people and it had begun raining. We only had another couple of hours of daylight left; it was time to get moving.

"Sahar, where are the fish people?" I said.

"I do not know, they have not answered."

"Well, how far is it to their camp, walking?"

"Half-an-hour, maybe slightly more."

"Well, what are we waiting for then? Send one of the men now," I said. Sahar looked a little affronted by this command. Debbie had warned me he was a little sensitive and we would have to work to gain his trust. But I also had to assert my authority as the leader of the expedition. It was important for him to understand that we were serious about this undertaking.

He obeyed and within forty-five minutes, the guide named John Kennedy returned via canoe with two of the fish people. I decided that Keith and the other guides would go in the first draft so they could begin building the shelters and Keith could have maximum rest. I would go in the second run with Andy, Sahar, and the remaining kit. The canoes are rough constructions and, like those of the Congo, hand made from a single tree trunk, it seems.

John, one of the fishermen, helped me haul my kit on board his canoe.

"Hello, Adam, pleased to meet you. I speak good English," he said.

"Thank you, John. Pleased to meet you, too," I replied.

"I am called John, John Kennedy, after the U.S. president," he said proudly.

"That seems a good choice."

"Yes, a very good one," he grinned, flashing a perfect set of large white teeth.

Arriving at the small jetty, I bounded onto the elevation and started to chop up wood to help the guys who were building the shelter. As I've said, I learned from my previous experience in the Congo that we would sleep in the shelters that the locals built; there would be no artificial separation between us. In little time at all, we had built the shelter, unpacked our gear, and set up a fire so that we could eat as soon as possible. Andy put on a big pot of smash loaded with bacon bits; after the day's exertions, it tasted fantastic. Sahar lit the paraffin lamps and we all settled down for the evening.

Keith assembled a contraption we lovingly called '"the crap robot," basically some sort of mosquito net with wires around and within it like a giant caterpillar. The guides loved it and laughed for hours over it. Keith entertained them by shining a flashlight in his mouth and growling. It was a good release after the day's rigors and helped us to bond as a team.

That night a severe storm battered our construction. Howling winds and lashing rain whipped the roof off and we were forced to make repairs. Keith's robot went scooting down the hill, and off he went after it. The sight of him grappling with a giant white caterpillar in the jungle is not something I will soon forget.

"Well, we might not find the yeti, but 'Sir Keith' there has already slain a monster," said Andy.

I noticed that night, that unlike the Africans, the Indonesians become nervous when the lights went out. They are expert guides and great survivalists, but they are not native and relaxed in the jungle, like, say, Sam was in the Congo. They behave very much like the military do – living *through* the jungle rather than *being a part of it*.

In the morning, Suaraham cooked up some garlic noodles for breakfast. After breakfast and the necessary ablutions, I decided to split the teams up and allocate guides. Sahar would come out with Andy and me, and Sansung would lead Keith on his first jungle trek.

The plan was to move to high observation points that covered a three hundred and sixty degree angle, wait for an hour in complete silence, and then moved on to the next positive locale. We would do this for several days in each camp. Sahar had outlined three or four possible camps that he knew, which correlated with known sightings of the *orang-pendek*. We would also hack a base in unexplored jungle to see if there might be any evidence of the creatures in areas not traversed by man. So we would not just stick to existing pathways. If a lead developed, we would plough on with it, regardless of where it took us.

I quickly learned just how damn good a tracker Sahar was. Just by the smallest scratches, or broken twigs, he could identify tiger, bear, or tapir – indeed, any of the creatures there. On the first day, we found some unusual tracks.

"I cannot explain these tracks, Adam," said Sahar. "I am very puzzled by them." Andy and I looked at each other, slightly bemused. From a tree, Sahar also pulled off some hair, using metal clippers to do so. "And I do not recognize this either – maybe, maybe it is *orang-pendek*, but I cannot be sure."

I took the clippers from Sahar and placed the hair in a sealed plastic container. "We'll get it analyzed," I said to Andy.

On the way back after our day's exertions, we discussed finding the hair. "Ad, stop me if you think I'm talking crap, but it seems like we got lucky too quickly," said Andy.

"Yeah, I agree," I said. "Maybe it's just beginner's luck – who knows?"

We then went back to camp to find Keith complaining about being stung by those "fucking nettles" in the "nasty horrible jungle." Keith and Samsung had found nothing.

John Kennedy had brought fresh fish in return for rice and Samsung mixed this in with chilies. Although this tastes okay because the fish are small, when chopped up the head remains intact so it looks like you are eating a bowl of fish heads.

"A delicious fish head supper – what more could I ask for on vacation?" said Keith.

"Try an eyeball, Keith," I said, holding one up on my fork. "They really are delicious."

"You are a very sick man," replied Keith.

"No, he's a bloody animal," said Andy. "You should see the way he moves through the jungle. He swings his arms and crashes through like an orang!"

"I told you on the first day that he was a throwback," said Keith.

That night we brought out the English/Indonesian dictionaries. I decided I'd give the guys English lessons every night from then on for an hour or so. I enjoyed it and it was good to bond with the guys and give them something back for their efforts. For by now, mutual trust was developing and we were doing well as a team. The guys were extremely enthusiastic and came out regularly for their English lessons.

That night we heard large animals crashing about in the jungle very close to our shelter. I also heard Keith get up several times. Turns out he had it coming out from both ends. Though he stayed at camp that day, as Andy and I trundled off into the jungle, his situation did not improve. We had some tough decisions to make: if Keith's situation got any worse, we would have to stretcher him down the mountain. But late that night, Keith began to take in some porridge, which was an encouraging sign, and color began returning to his cheeks.

The following morning Keith was feeling slightly better and my initial concerns began to fade. "The best thing," I told Keith, "is for us to finish trekking here today. Tomorrow morning, we'll break camp and John will take you back to Sungai Penua. You'll rest there, and John will return to let us know you're safe and well." Keith agreed to the plan.

That day Andy and I had one highly amusing experience, which helped to lightened our mood. We were half way up a ridge when we caught a musty smell in the air. Sahar stopped us right away. By now we had learned that tigers are often followed by clouds of small black flies. Sure enough, the flies were clearly visible. Then, from behind Andy came the loudest roar I've ever heard. Andy looked at me perplexed and I shot a look back at Sahar, who looked perplexed too, though he was normally so confident in the jungle.

"Maybe he not happy. Maybe we should go now," said Sahar.

"Maybe we should wait and see him," I said.

"No, Adam, he is angry. Maybe we should go now," said Sahar, this time expressing genuine concern. And off we went, running back to camp.

I was looking forward to pushing off to camp 2 the next day. Camp 1 saw visits by the odd tourist and local, so in some ways it just didn't feel wild enough for me. Even though there was a strong possibility we'd found tracks and possibly even a hair sample, we had been all over the area by now and found nothing else, so it was time to push on.

The following morning we broke camp and say our goodbyes to Keith. Andy, Sahar, and I loaded up our packs and trawled off. I was sorry to see Keith go; he was always optimistic despite his difficulties and had tried his best. But the journey wasn't going to get any easier. It would be just over a week before we were due to see him again. John Kennedy would join us at camp 2, which was a little over six kilometers away, in three days time.

The path to camp 2 was fairly clear, but the humidity was pretty high so it made the three-hour trek quite an effort. When we approached the open shoreline that was to be our second base camp, Simian monkeys in the trees behind us howled warning cries to their compatriots. A thin mist crawled its way across the middle of the ridge behind us, and above it, a fish eagle flew in tight circles, riding thermals as it hunted its prey.

"This is more like it!" said Andy, flinging down his pack. This was true wilderness.

That night we lit a fire and had a few swigs of the vodka we had brought with us. At this elevation the sky is alive at night. You don't get that pink hue you see in cities, where streetlights pollute the sky. Here the stars stand out proudly against the sky, and gorgeous shooting stars are often visible.

"Make a wish," said Andy, as one crossed the horizon before spinning off into oblivion.

"I already have," I replied.

Our shelter at camp 2 was a magnificent construction, and this time we had the benefit of soft grass to sleep on.

The following morning, I awoke before Andy and went outside to find Pendi clutching a black and white snake. It was dead. "Look, Adam," he said.

"Give it here mate, I've an idea for that," I said, grabbing if from Pendi. I placed it gently on the top of Andy's sleeping bag and told all the guys to keep still.

Lying down, I shook Andy slightly. "Andy," I said. "I think there's something in here."

"What? Oh, fuck!" he said jumping up. "Shit!" He began dancing around and Pendi joined in.

"We dance, yeah!" said Pendi, imitating and exaggerating Andy's movements.

Eventually Andy realized the snake was dead. "You bastard! Where did you get that from?"

"It fell out of your pants during the night," I said, solemnly. "I regret to inform you that you have severe cock rot and are, in fact, a eunuch now."

"Very funny," he said. "You just wait, I'll get you back for that."

By now the smell of garlic noodles filled our nostrils and over breakfast we planned the day's trekking with Sahar.

At one point Andy turned towards me. "Ad?"

"Yeah?" I said.

"You know that snake. Where's it gone now then?"

We both looked at our bowls. I looked at Sudaraham, the cook. He grinned back at me.

"I don't know," I said to Andy. "Probably best not to ask."

That day we trekked up along the ridges. The paths were more slippery and overgrown here, and Sahar and Sansung had to hack their way through at times. Being a snowboarder, Andy's sense of balance was superior to my own, and this was to cost me as I slipped from a root and hit a path of stinging nettles. Now I knew a little how Keith felt.

Two hours into the trek, Sahar and Sansung stopped dead. They had discovered tracks. But this time, there was none of the doubt in Sahar's voice that had been evident the first time he had spoken. Crouching down, he studied the tracks closely and then conferred with Sansung.

"Adam, I have not seen this creature before. It's tracks, I do not know. This is the *orang-pendek*." He sounded surprisingly confident.

No sooner had he said this that we heard something crashing through the jungle about one hundred meters in front of us. "*Orang-pendek*," we all shouted simultaneously, and bolted off. We were running now, half stumbling on roots as we went. The crashing continued. We rushed ahead until our lungs burst and all we could hear was the sound of our own breathing. The crashing had gone.

Whatever it was, we'd lost it.

We returned to the spot where we'd found the tracks.

"Can we gip it, Sahar?" I asked. Gip is Plaster of Paris.

"No, I don't think so, Adam. I don't think the gip will take," he said disappointedly.

Looking around the area, it was clear that this creature didn't move surreptitiously at all. Tree trunks had been rolled over as it looked for insects and there was evidence of discarded fruit. I recalled an observation Debbie had made when she had seen it: "It seemed to move through the jungle snacking, almost treating it like a giant supermarket." I could see what she meant.

Sahar measured the print and marked its location on the map. It had four toes and a thumb, eleven centimeters across, fifteen centimeters from the tip of the toe to the base of the heel. The heel itself was six centimeters across.

We were delighted; it had been a great day's trekking, but we still hadn't got enough. Discussing the route with Sahar, we'd decided we'd spend another day or so here, and then trek through some of the uncharted jungle between Bantu Mesjid and Ulutebokanam. The plan was to climb Tebo Kanan, a huge mountain that filled the skyline in front of us.

What if the creatures were moving progressively round the lake like us? I wondered.

That night we had another English lesson and then off we went to sleep.

The next morning, we were all really pleased by John Kennedy's surprise early arrival. He'd made good time by getting one of the fish people to row him across the lake. He and his son joined us, bringing with them a scrawny old chicken. John told us that Keith was okay, though he had managed to fall and cut his arm on the way back.

"He'll be looking for a burger in Sungai Penuh as we speak," said Andy.

"Well he won't find one," I said. At least we didn't have to worry about Keith's health.

After consultation, I decided that Sahar, John, Andy, and I would set off for the area where we had found the tracks the day before and make a fresh effort to find clearer ones. When we got there a few hours later, John searched the undergrowth vigorously.

"Sahar," I said, "I want you to gip that print anyway, even though it might have degraded. We've got nothing to lose; it's worth a shot."

"Okay, Adam," replied Sahar. Suddenly we heard John shouting excitedly. Slightly down from the first track and obscured by thick undergrowth was a pristine track.

"Will it work?" I asked.

"Yes, I think so," said Sahar. He always chewed over everything carefully.

"We'll know in half-an-hour," said Andy.

Sahar seemed more confident than ever with his find. Sahar and John poured the thick paste into the earth. Andy and I chewed on Pepparamis.

"I think I've had enough Pepparami's to last a lifetime," I said.

"Yeah, or until the next expedition," replied Andy.

We then waited for the gip to harden. It was a long half-hour. In the meantime, we did find more hairs, which we bottled.

Gradually, Sahar began lifting the mold of the degraded print from the ground. It was not that clear, but toes were distinguishable. When we removed the mold from the second print, however, we all sent up a simultaneous cheer. The print was perfect!

Like kids, we rushed back to camp determined to show off our new plaything to the rest of the expedition. Sansul had wrapped himself in a shawl so that only his whiskers and eyes were visible. "After another couple of weeks in the jungle, I could almost fancy you in that!" I said, laughing.

We had a great feeling of triumph that day. The fisherman joined in and pulling his chicken on a string, invited it to do some bizarre dance with him. A dispassionate observer arriving on the scene would have had us all committed – chicken and all.

The fishermen stayed awake all night talking about the *orang-pendek* and our find. They were as excited as we were. We decided that for camp 3, rather than walk the whole way round (five hours) and then spend time skirting the shoreline and building camp (minimum two to three hours), we'd use the fishermen to take us. This would also allow us to skirt the shoreline and look for places to camp, rather than doing it with all our kit.

Our guide across the lake was a fisherman who had seen the creature. When we stopped where he had had the sighting, he began thrashing his arms around. "It was here, look, look at the fruit," he said. There were bite marks in the fruit. But the undergrowth was very thick here, and we saw no tracks, so we moved on.

Eventually we arrived at where the others had made camp 3. We were not greeted by their usual cheery wave. In fact, they all looked thoroughly miserable.

"What's the matter?" I said.

"The bees, they are very bad here. John has been stung," said Sahar.

John looked like he had a small egg on the side of his head. He looked at me with strained, half closed eyes. I handed him the anti-inflammatory cream. He put a large dollop on his egg. We stayed for about an hour or so waiting for the fish people to get back.

For some reason, perhaps because no people come to this side of the lake, the insects here seemed at least twice the size of those we had encountered on the other side. We'd obviously camped near a high concentration of them. I decided that the others would make camp further on, using the boats to get there. I got Pendi to call the fishermen back. Much of the land around the lake here was impassible, so Andy, Sahar, and I would wade in the shallows of the lake and then cut our way across to camp. If we waited for the boats, it would've been nightfall, and we wouldn't have made it.

When the boats arrived, the others set off, and we began our wading across what proved to be some bloody sharp stones. Cutting your way through virgin jungle is tiring and hard work. Andy picked up a gash on his forehead, which I dabbed with surgical spirit, but apart from that, we arrived in one piece at camp 3. Away from the bees, morale was hugely improved and everyone was in good spirits.

Andy looked up at the mountain behind us and said: "Tomorrow, mate, we'll climb that. Ready?"

"Ready," I replied crisply.

After the English lesson, we settled down to sleep, but again our sleep was interrupted, this time by a visitor. Sleeping in the jungle, you get used to insects crawling over you; you either accept it or you'll never sleep. However, at 2 am I felt a dead weight on my sleeping bag. I leaped up and saw something fly into the bushes.

"What was that?" I asked Andy.

"I dunno," said Andy. "But it jumped on me first. Naturally I had to flick it onto you to get you back for the snake."

"Cheers!" I said.

Our "visitor" returned several times. Looking it up in the guidebook, I reckoned it was an inquisitive "moon rat." It could've been worse, I suppose.

The following morning, we set off for the mountain with Sansung and Sahar.

The trek up through Gunung Kulu to Tebo Kanan was extremely arduous. At times, the slopes were eighty percent in front of us, with dubious branches being the only footing keeping us from falling hundreds of feet below.

It was during this climb that I realized just how British we were. On several occasions Andy and I would remark on an orchid or flower we had seen: "Oh, look at that, what a magnificent hue of pink."

Sansung would remark "No!" and shake his head before being egged on by Sahar. Of course, Andy and I were just as nervous as he; we were just demonstrating our stiff upper lip, while the Indonesians, freer with their emotions, were telling it like it really was.

Climbing though the mist, we eventually reached the top, totally knackered. We sat down to devour our last remaining Mars bars.

"Look, Ad, the packets are inflated," said Andy, holding one.

"Yeah, because of the altitude," I remarked.

By the time we arrived back at the disassembled camp, it was nearly dark. As the boats arrived, I had the others take them to camp 4. Sahar, Andy, and I waited for them to return.

"Adam, I am very proud. What we have done today, no locals, no tourists do. We have done very well," said Sahar. He was beaming enthusiastically.

"It seems like the guys have enjoyed themselves as much as we have," said Andy.

"Yeah, we've become a really strong team," I remarked to Andy.

By the time the boats returned it was nearly pitch black. I jumped in the boat with John Kennedy. Andy and Sahar went in with one of the fishermen. John was so tired from all the rowing that he asked me to help him, which I did. In a single day, we had hacked through jungle, climbed a mountain, and canoed five kilometers across a lake.

After a meal and the obligatory English lessons, Andy asked: "What do you call a Yeti with no bum?"

"I don't know," I said.

"The unbummable snowman!" he quipped.

"Oh dear, it really is time to leave the jungle," I said. And it was. We were getting near the end of our expedition.

The next morning we awoke for our final trek. John and Sudaraham were frying some noodles and fish again. John Kennedy had caught fresh fish, and without waiting for it to be killed, plopped it straight in the hot oil of the cooking pot.

"I'm not keen at all on that," said Andy. "You should kill the poor sod first."

"I agree," I said. "Joy, I think I'll just have buttered toast this morning!"

"Croissants for me!" Andy exclaimed.

John, who by now realized we were being tongue in cheek, said, "When we get home!"

The last day's trek was, I admit, fairly half-hearted. We saw little apart from areas where Jeremy Holden had placed his camera traps as part of Project Orang-Pendek. That evening we met Eddie, the guide from Sungai-Penuh. He had brought with him a Belgian biologist named Phillippe who was studying spiders. Phillippe, who was spending a few days in the jungle, joined us for a meal and filled us in on world news since September 11th.

The following morning, we packed up camp and prepared to set off down the mountain. The journey across to the point of descent threw just one more hazard at us. Pendi, bounding across a log, held out a branch to help me cross. On one side was still water, on the other was a waterfall cascading thousands of feet to the ground below. I made the crossing and turned to Andy.

"Cross on the side of the log in the shallow water, and hold the fucking branch tightly," I shouted. "Otherwise you could go over the waterfall."

"I don't need any encouragement," said an understandably nervous Andy.

Five minutes later we had all made it safely across. Climbing the hill, I took a last look at the lake. My friend the fish eagle continued to circle, as he and his kind had done for thousands of years.

We descended fairly quickly, but when I saw the strain on Andy's face, I decided to make plenty of rest stops. Indeed, it was the only time that Andy ever asked me to rest. He confessed later that he had been at the point of exhaustion. We'd trekked for weeks now, climbing ridges with little sleep, and it had taken its toll. When I looked at the photographs later, he looked ten years older at the end of the trip than he did at the beginning. He had given his all and had been a great companion. If he looked knackered, then, I reasoned, I mustn't be far behind.

When we arrived at the camp entrance, we stopped for some team photos and then headed off to Sahar's. Once there, we all went off for a bath in the river. It was the weirdest feeling, walking on flat ground after two weeks. The air was also noticeably thicker and warmer. That night, we each had a bed for the first time in weeks.

The next day we got on a bus to get back to Debbie's. Indonesian buses defy the laws of physics. Basically, they operate on this principle: how many people and their bulky belongings can we cram into a little mini bus? The answer in our case was sixteen! At one point, another driver flagged up down. I asked Sahar what was going on. "Oh, the police are up there taking bribes. We will go the other way."

Eventually, we arrived back at Debbie's. Neither she nor Keith was there. Jeremy and a researcher named Matt were there.

"Well, how did you go on?" asked Jeremy earnestly before I'd even had time to introduce myself.

"Very well," I said. We pulled the prints from out of the bag. Both he and Matt inspected them.

"Fucking amazing. Well done! Well done!" Jeremy, who had been a little muted when we'd first arrived, suddenly sprang to life, delight evident on his face. "Listen guys, Keith isn't coming until three, come and have lunch with us, you've got to show this to Frank!"

"Who's Frank?" I asked.

"Frank Lambert, he's an ornithologist. He's been umming and ahhing over whether he believes the creature exists. Wait till he sees this."

Jeremy must have been the tallest person in Indonesia, I reflected as we climbed into the jeep. He was easily six feet three inches, if not more. Over lunch Jeremy recounted how he'd seen the creature on Gunung Kerinchi. He had followed it into a ditch, but it had disappeared over the rim. "I've spent five years since then looking for it, with no success," he said. "If only I'd photographed it at the time…" His voice trailed away. "We've seen no real evidence of them since." Unlike Debbie, Jeremy reckoned the creature was a species of gibbon.

As I spoke with Jeremy, I had the feeling that he was trapped here, almost against his will. After he'd seen the creature, it was like he couldn't leave until he'd proved it was there. His frustration was very evident. Frank rolled in, wearing the brightest Hawaiian shirt I'd ever seen. Jeremy introduced us and I pulled the footprint from the bag. Frank examined it closely. We waited. None of us made a sound.

After an eternity, Frank spoke. "Fuck! You really do have something here!"

Jeremy laughed. "You see! You see!" he said.

We had a good chat with Frank and arranged to meet up with him for a beer the following night as we were all going to be in Pedang. When we returned to the house, Keith was waiting for us. We were delighted to see him and he gave warm handshakes to all the guides as well. The guides then collected their money. As we paid them, I thanked each one individually. At the end, we had a toast – with coffee, naturally, since they are all Muslim.

"Once again," I said, "I want to thank you all so much for what you have done for us – brilliant and exceptional are two words I could use, but I think a third, 'friends' is more appropriate, because that is what you have become to us."

And so they left.

Andy and I then went off with Keith to hear about his adventures. Keith had had a fairly quiet week. He'd had lunch with Debbie a few times and been to the hot water springs. But around Sungai Penuh, he was a local celebrity. As we walked through the streets, people waved and cheered, "Mr. Keith, Mr. Keith." He'd also been teaching in a local school, and after a week in the town, many people knew him. Keith loved his celebrity status.

"Tonight, boys, I am taking you for a treat," he grinned. "Tonight, Ad, you'll be having a burger!"

"A burger?"

"I have found a chicken house!" He punched the air triumphantly. And indeed he had, for as we turned the corner, a red neon sign proudly announced "Sungai Penuh Fried Chicken." "It's not quite Kentucky Fried, but it'll do," said Keith. After all, they sold burgers.

As we walked in, the assistants all shouted, "Mr. Keith, Mr. Keith." There was no doubt that Keith was the king of the chicken house. He strode purposefully to the counter and introduced us to his new friends.

"For many of the locals, a meal here would be a couple of weeks wages," said Keith. "These guys earn very little money, so I've hired a couple of them to give me tours of the town while you were away." By this point, Andy and I were barely listening; we were devoured our burgers like ravenous animals. Within a few seconds, they were gone.

After the meal, we headed off to meet Jeremy and Matt at the club, where we order some beers, enjoyed the karaoke, and Keith danced with some of the local girls.

The following morning we packed our gear and waited for Debbie and the driver Armin to arrive. When Debbie showed up, the first thing she said was: "C'mon, c'mon let's see it then." So I unwrapped the footprint and handed it to her. "Oh, marvelous," she said, "this is simply marvelous! You've done so well, boys. I'm so pleased. I feel so vindicated." She looked close to tears.

Eventually it was time to say our goodbyes. Debbie and Matt waved us off and we promised to keep in touch. In the minibus, Armin had changed his tape from Bob Marley to Rod Stewart, so Rod duly sung for the next seven hours as we wound around the dodgy coastal road to Pedang. We were only there for one night and had just enough Rupiahs for separate rooms. Of course, we met Frank for a few beers later on in the evening at the hotel disco.

While the others whipped the police chief in a game of pool, I chatted with Frank.

"Y'know things are a bit tense in town with all the trouble in Afghanistan," he said.

"How's that?" I asked.

"Well, only last night some extremists came to the hotel and asked if any Americans were staying. When the receptionists confirmed that there was one, these guys told them to tell him that he had one night to leave – or else!" Frank drew his fingers across his throat.

The next day, we headed off to Singapore. It was time to put our feet up and enjoy life – everything from Singapore Slings at Raffles, to drinking seven pints in an hour at Hooters, to Chilli crab at Fatty's, and high tea at the Regency.

It had been a great expedition.

Chapter Nine

Back To Sumatra

Our footprint casts and hair samples made world news. Both the national and international press were fascinated by the idea of a "jungle yeti." Even more important, though, we began to get credible scientific backing. This obviously has to be the ultimate purpose of any field research. To gather information such as I do, you don't have to be a scientist, although an understanding of the best way to gather samples obviously helps. What's most important is that whatever evidence you collect must be independently tested by credible scientists. It's all very well to say, "I've seen one, and I know what I've seen," but that has absolutely no scientific value.

I was happy to have any professional analyze what I found with one caveat: if they concluded that what I'd found pointed to the existence of an unknown species of primate, then they should have the guts to say so publicly. If they disagreed, then fair enough, that would be the end of the matter.

With the help of Loren Coleman, I sent photographs of the footprint to a number of reputable sources, among them Colin Graves, Karl Shuker, and David Chivers, a primatologist at Cambridge University. We made a copy of the cast for Chivers and Andy took it down to Cambridge so that he could examine it. Chivers had no doubt that this footprint belonged to an unknown primate: "These footprints were very exciting, very unusual because they were of mixed character from all the different apes and humans. They have the toes that are shorter, more like humans, the heel is like nothing, in that it's curved. We call it banana foot."

Hans Brunner of Deakin University, near Melbourne, analyzed the hair samples. Brunner had been instrumental in his analysis of the hair samples in the famous Lindy Chamberlain dingo baby case in the early 1980s. If his opinions were good enough for a murder trial, then they were good enough for me. Hans has spent ages gathering hair samples from indigenous species as well as hair from humans and other primates such as chimpanzees. After extensive and comprehensive research, he came to the bold conclusion that the hair we had found did indeed come from an unknown species of primate.

Unfortunately, when we sent the hair off for DNA analysis for a *National Geographic* documentary on the *orang-pendek*, Todd Disotell, who conducted the analysis, was unable to extract primate DNA from it. There was a human trace, as most likely we contaminated the sample in the collection process. We were too blasé about it, just sticking the sample in Andy's camera film container. I take responsibility for this; you live and learn. I respect Todd and thanked him for his efforts. I now take a sample pack that Todd gave me so I won't make the same mistake twice.

When I traveled to Australia the following year to see a friend of mine who had immigrated there, I took the opportunity to meet Hans Brunner. (I also took the time to trek in the Grampians where the Tasmanian Tiger, which is thought to have been extinct since the 1720s, had been seen and might still be lurking.) I was enormously grateful to him, as I was given to understand that some of his colleagues had warned him against the topic, saying that he would leave himself open to ridicule if he were to declare publicly that an unknown primate existed. Yet he had the personal integrity to say so because he believed it was true.

As an aside, I was told by a respected and eminent scientist, who shall remain nameless, as it was said in confidence, that if we had been talking about a new species of frog, there would have been no difficulty in publishing scientific papers about it. But because this was about an unknown, human-like primate, the burden of proof was going to be that much higher. I was annoyed at his remark, but I understood what he meant. Most scientific discovery is made despite of mainstream opinion, not because of it.

Like a dog chasing its own tail, the more scientific backing we received, the more publicity we received as well, everything from the *Richard and Judy Show* to *Channel 5 News*. Keith basked in his temporary fame from having full-page pictures of himself in the newspaper, while Andy was frustrated that nobody ever spelled his name correctly.

Encouraged by our success in Sumatra, we resolved that we'd go again in 2004. I had one main objective this time: I wanted to see an *orang-pendek*. I knew it existed and I'd done my best to prove it existed, but what I really wanted to do was to see one for myself.

But when we reached the forest above Buttinghi in 2004, I was shocked at just how much de-forestation was going on. The primary jungle is gone; it's gone forever. The *orang-pendek* lives in primary jungle. Aside from its infinite beauty, this jungle was home to a number of unique and marvelous creatures. As I stood on the farmland on the edge of the National Park, I could see the shriveled out skeletons of the trees. Time seems to have run out for the *orang-pendek* and it filled me with a desperate sadness. When we met up again with Sahar, we found more footprints in the forests above Buttinghi Village, which is just a few kilometers from Gunung Tujuh. Apparently the locals had seen an *orang-pendek* eating fruit on the edge of the farmland there. I staked out the farmland, but to no avail; it had already moved on.

In 2007 I was asked to return to Sumatra by the *History Channel* for a documentary they were planning on the *orang-pendek*. This time I would team up with Jeremy Holden.

I met the film crew – producer Jared Christie and cameraman Aaron Achtenberg – in Singapore. We joined Jeremy in Padang and met our chief guides, Sahar and Doni, in their village at the foot of the volcano. For most the expedition, we split into two teams. Sahar and I were the tracking team, picking up any trails we could find of the *orang-pendek*, while Jeremy and the guide Doni set up camera traps.

It was going to be tough doing it all in the limited time we had available – and Jeremy and I nearly didn't make it over the lake on the way in. We were delayed at the National Park try to obtain the necessary film permits, and as a result, by the time we started to cross the lake at the top of Gunung Tujuh, it was getting dark. We would have no choice but to traverse the lake at night. We divided everyone and the gear into the various pirogues required to traverse the lake, hiring the local fishermen to do the rowing.

Jeremy and I were the last to leave. As it grew dark, to our surprise the fishermen took us increasingly far from the shore. It was at that point I noticed that there was a hole in our boat and that we were about to sink.

"We are going in," I said to Jeremy. His eyes widened and I saw him grip his camera like a baby. He ordered the fishermen to paddle desperately for shore, and as we did so Jeremy frantically baled with a pan. We made it to the shore with seconds to spare. I had narrowly avoided a potential drowning.

Would I have made it? Probably, as I'm a pretty good swimmer. (I won competitions as a child). Jeremy, by his own admission, probably wouldn't have. He's not the strongest of swimmers and he was understandably worried about hanging on to his camera. All the same, we were both glad to get to shore.

Over the next few days we got down to our respective tasks. I was fascinated by the "musk tags" from the Osmick Record Corporation that Jared had given me. Supposedly, they were designed to attract the attention of the Sasquatch. I had my doubts, but nothing ventured, nothing gained, so I placed them on the pathways at the side of the jungle. I've saved a few and hope to try it on Cameron Diaz someday.

Sahar and I picked up plenty of animal trails – deer, tiger, bear, etc. And again I picked up the trail of the *orang-pendek*. I had theorized that the creature had a specific range and, that as an opportunistic feeder, it would probably follow familiar routes in search of food. This proved to be correct, as I found new prints near the area where I'd previously found prints in 2001. The *History Channel* filmed the evidence. From the clearly identifiable trail we made two excellent prints, which were analyzed by anthropologist Jeff Meldrum of Idaho State University for the documentary. While he didn't doubt the veracity of the evidence I'd produced in 2001, his analysis concluded that the prints we found in 2007 were bear prints unfortunately. Though I've found credible evidence to convince respected scientists, I concede more still needs to be discovered.

Being back in Sumatra searching for the *orang-pendek* made me very emotional. While I kept finding evidence of the existence of this wonderful creature, the amazingly beautiful jungle it called home was being chipped away by the greedy, and soon it would all be gone. Was this the same animal I'd found in 2001? How many were left? Was this one of the last ones walking a lonely trail to extinction? Even if I did prove its existence, would that help? Lowland Gorillas may well beg to differ that publicity helped their cause.

That was the strangest day of my life. I swung between happiness and pathos, between hope and fear. Jeremy is intrinsically pessimistic about its fate. I hope he's wrong.

On the last night before we left, an earthquake struck. The whole ground seemed to turn to liquid before my eyes and the birds and gibbons went mad. Sahar grabbed a tree and began crying; he worried desperately about his family in the village down below us. We sent him and Doni down right away, of course, to get news of his family, and Jeremy, who could speak good Indonesian, went with them. Aaron, Jared, and I spent the night in the jungle, and with little else to do we drank whisky and listened to Amy Winehouse.

I had a broken sleep that night in the homemade shelter while the others slept in tents. One particularly large animal came right up to where I was sleeping. I heard it crash around and lap the water. But I couldn't see what it was, as I didn't turn over until it had gone.

The following morning the fishermen returned to pick us up. The good news was that the damage from the quake was not as great as we had feared. Sahar, Doni, and the villagers and their families were okay. Generally across the islands the damage was not extensive but people had understandably been scared. We saw people sleeping in tents and we all experienced several tremors. There had been another earthquake in the morning and there were repeated aftermaths over the next few days.

I saw several collapsed buildings in Padang when we returned there, and our hotel was covered in structural cracks. We did all return in one piece though, and we parted with Jeremy at Pedang, the airport having reopened after a temporary closure during the quake. I left the Americans in Singapore and continued to Manchester on my own.

I really hope the *orang-pendek* can be saved.

Chapter Ten

The Mongolian Deathworm

Between my expeditions to Sumatra, I traveled to Mongolia on two occasions, the first time to look for the *Allghoi Khorkhoi*, the Mongolian Death Worm.

A Death Worm can't be real, can it? This five-foot long creature, which is said to live in the deserts of the Gobi, supposedly can kill people at a distance with lightning. I was particularly alarmed by this description.

The first reference to this remarkable Death Worm in English appears in Roy Chapman Andrews' book, *On The Trail of Ancient Man*, published in 1926. Professor Andrews, apparently the inspiration for the Indiana Jones movie character, was asked to investigate the account by the Mongolian president.

Czech explorer Ivan Mackerle, who has done the most extensive searches for the creature, provides the best description to date: "Sausage-like worm over half a meter (20 inches) long, and thick as a man's arm, resembling the intestine of cattle. Its tail is short, as [if] it were cut off, but not tapered. It is difficult to tell its head from its tail because it has no visible eyes, nostrils or mouth. Its color is dark red, like blood or salami… It moves in odd ways – either it rolls around or squirms sideways, sweeping its way about. It lives in desolate sand dunes and in the hot valleys of the Gobi desert with saxaul plants underground. It is possible to see it only during the hottest months of the year, June and July; later it burrows into the sand and sleeps. It gets out on the ground mainly after the rain, when the ground is wet. It is dangerous, because it can kill people and animals instantly at a range of several meters."

The Death Worm does seem like something out of Hollywood, more than anything else (e.g. *Tremors*). Furthermore, a "worm" couldn't survive in the harsh environment of the Gobi – simply put, the temperatures would dry it out. Putting such doubts aside, however, there have been stories about it for centuries.

I was fascinated by the idea of such a creature and thought it worth looking into. After contacting the Czech explorer Ivan Mackerle through a lead given me by Karl Shuker, Andy and I discussed the practicalities. We spent the best part of a year researching the terrain, the equipment we would need, and most importantly, the sort of people we'd need to assist us in Mongolia. Eventually, based on Mackerle's extensive local experience of the Southern Gobi, we decided on a team lead by Byambaa. We hired him as our "fixer" in the capital of Mongolia, where Byambaa informed us that we were to travel with a driver named Jagaa and a cook-con-interpreter named Bilgee.

As we were putting the finishing touches to our plans though, a disaster struck in the form of SARS, a respiratory disease in humans that is caused by a coronavirus. It began to spread all over Southeast Asia. We had booked and arranged our flights to Asia to go via Bejiing, which as our departure date approached, seemed to be in a lockdown mode. Few people were venturing out and those who did were scurrying about in masks; that was the perception from the TV coverage anyway. Andy and I discussed whether to go or not given the situation. But we pretty much agreed that we were going to let nothing stop us from making the trip. I've always thought that it's the cities that are the problem; they are full of people, regulations, and illness wherever you are in the world. Once you are away from them, into nature, then you are free.

Still we were shocked to be greeted by Air China stewardesses in masks and gowns as we boarded our flight from Heathrow to Bejiing. When we arrived in Bejiing some ten hours or so later and wandered into the city center, there was hardly a soul around. This was Bejiing, a vast city, a city seething with cars and yet there was only one little vehicle chugged unhindered through the deserted streets to our hotel, The Traders Bejiing.

When we arrived, we were required to have a mandatory check of our body temperature check, as was required of all guests. If our temperature was elevated, we'd have to report to the authorities. The concierge took mine.

"Your temperature is very low," he said frowning.

"Well, that's a good thing then, isn't it?" I said. He frowned but let me check in anyway.

We decided to do some sightseeing, and as our hotel was right near Tiananmen Square, wandered down to it. This vast square was virtually deserted. There was one solitary figure flying a kite. Andy and I wandered into the Forbidden City without queues of tourists; there was nobody around but us. I loved the Forbidden City. I had read avidly about it before I left. As I sat in the middle of the Emperor's Garden, I reflected on how this introspective and remarkably sophisticated culture had disappeared under a web of intrigue and foreign influences alike.

The Chinese were nothing if not very, very polite to us while we were there. By this time though, most foreigners who could leave the city had done so; we must have ranked as something of a curiosity. I don't know it for sure, but we both felt we were being watched. I'll give you an example. Sitting in our hotel room that night, Andy said to me, "I really fancy a drink, mate." Less than two minutes later, a waiter knocked at our door, asking if we'd like a drink. Nothing strange there you might think, just a coincidence, right?

In our room, were some Chinese pears, which were delicious. "I'd like another of these," said Andy, as he munched through his. Lo and behold, the steward was back at the door offering us some pears.

The following day we went to The Great Wall of China and the Ming Tombs where we were greeted to the unusual sight of Chinese soldiers marching in masks. I was quite used to the city now; once you got used to the creepiness of a deserted city (like *28 Days Later*), there are a number of advantages, like not waiting for anything and not being hassled on the street, by vendors selling tacky souvenirs, for example. But I suppose the downside was that you could catch a potentially fatal disease.

The following morning we set off for Mongolia, landing in Ulan Bator, where we were given the obligatory temperature check and were met at the airport by Byambaa.

That night we met up with Bilgee and, pouring over the map, outlined our plan. Simply put, we were to head to an area not far from the Chinese border where Mackerle had said there had been the greatest number of sightings of the Death Worm. We wanted to see if we could pick up any recent eyewitness reports. If we did, we would begin searching the area.

Bilgee agreed to the plan, as he sucked hard on the toothpick he had stuck in the corner of his mouth. "It will be hard being in the Gobi that long," he said.

"I'm sure we will be okay," I said.

"Yes, there will be no problem," he agreed finally, grinning from ear to ear.

The following morning we met up with Jagaa, our driver, and left the suburbs of Ulan Bator behind.

The first thing that strikes you about Mongolia is the sheer vastness and forbidding beauty of its plains. Endlessly rolling, punctuated by the occasional herd of antelope or a hurrying marmat, these plains are utterly unique. I loved them. Even now, when I look across fields in England, I think of those in Mongolia and get goose bumps.

Mongolians are, in my experience, the most hospitable and genuinely friendly people in the world. Those who live on the plains are nomadic, inhabiting felt tents called Gers and moving with their animals, primarily horses and goats, as and when it suits them. Nobody owns this land; it belongs to everyone. The concept of ownership of land is primarily confined to the cities, it seems. Vicious dogs guard most Gers and their cattle; indeed, a common shout among Mongolians when approaching a Ger is: "Hold the dogs!" The dogs are a deterrent to wolves. The dogs would regularly chase our jeep as we crossed the plains, and we used to bet on how long they would chase us as we passed their territory.

Before I went, I spent some time researching Mongolian customs, which are unique. For example, the passing of snuff is an important tradition when you exchange pleasantries. It is also customary to sit on the left hand side when you enter a Ger. I always made sure I went in first as the leader of the expedition. It is also very important not to touch a Mongolian's hat. They don't expect you to get everything right, but the fact that you are making an effort, I think, shows your host respect.

Anyone can enter a Ger and sit down and be entertained as their guests. Mongolians are a gregarious people. I think it is fair to say that they like a good conversation and a good slug of vodka. Both

Andy and I were amenable to both, so we got along well. What we weren't to keen on, though, was the food.

In winter the Mongolians will kill animals (e.g. sheep) and consume all edible parts. In the summer they produce dairy goods such as cheese and airag, which is normally fermented from the milk of a mare. Andy called what we ate "lucky dip soup" because you never knew what you were going to get. As the leader of the expedition, I sat at the top end of the Ger and used to get first dip. I always used to pray I'd not get an eyeball or a bullock. But whatever it was, I ate. It would have been the height of bad manners to refuse and, after all, I'm British.

On the way to our camp we stopped off near an abandoned monastery. It was a sobering thought that just ten years earlier the communists had murdered all the monks who lived there. It was at this time that the Czech explorer Ivan Mackerle had first visited the area; perhaps the reputation of Mongolians being tight-lipped and reluctant to talk to foreigners stemmed from the fact that their society had been so consistently repressed since the Stalinist era.

While I sat in the camp at night, pondering such questions, a small snake slithered past, arching its head towards me as it did so. "Wolf snake," said Jagaa. "Very poisonous."

Gradually, as we traveled on, the terrain became less verdant, the green turning to brown. I did take every opportunity while we rested at the Gers to talk to as many Mongolians as possible about the Death Worm. While most had heard of it, no one was able to provide firm clues to its habitat or location.

It wasn't until we reached the Gobi National Park that we got our first substantive lead. Speaking to one of the park guides, I learned that a local man had apparently made a carving of a Death Worm. It now sat proudly in the museum among the snow leopards and mountain goats. The guide also knew of an elderly man who had spent many years researching stories of the Khorkhoi. He lived in a tourist Ger about thirty kilometers (nineteen miles) away.

After a day tracking, we went to visit him. He pinpointed precise locations on our map where our search might prove fruitful. He explained that the creature appeared mainly in June or July, that it burrowed in the sand, and that the most common sightings occurred just after a rainfall and when the Goyo plant (a parasitic plant with a yellow flower) was in bloom.

One particular piece of information did give us cause for concern. The valleys, the old man remarked, were home to a particularly vicious spider that wasn't averse to running and charging at human beings! I was a little skeptical of this, as in my own experience, spiders, snakes, and other creatures we often invest with threatening qualities normally disappear at the first sign of trouble. But Andy, who counted the spider as his least favorite creature, looked distinctly worried.

The following day we headed off to a region never before traversed by foreigners and picked up our first really good eyewitness lead – a young man who, only three years before, had seen a Khorkhoi near a well in the area. It turned out that several people in the local villages had also seen the Khorkhoi and we decided to interview them before traveling on to the next promising location.

We were told that near the local army base lived a very reliable witness, a man by the name of Khurvoo, who had seen the creature on three occasions. Upon locating him, Khurvoo was very animated in his description of the creature, how it had moved, and how shocked he had been on seeing it the first time.

All of a sudden though, a soldier arrested us and placed us under guard. As Andy had been sitting in the front of the jeep to film, they took him for interrogation first, then Bilgee. I sat frustrated in the jeep, wondering what was going on. A few hours later, it was my turn. I was escorted into the base accompanied by an armed guard. My mind whirred. I began memorizing the army base's layout and my route in, contemplating an escape if it became necessary. I didn't think the chances were great though, but if it came down to it, I'd try. I'd seen soldiers with Alsatians and heard the buzz of a small light aircraft overhead.

I was shown to a small room where Bilgee sat on one side of a desk. On the other side sat two officers, a major and a colonel. The colonel began asking me questions.

"Are you a Chinese spy?" he asked. "Tell him that if he is a Chinese spy, we will kill him!"

"Do I look like a Chinese spy?" I replied.

"What are you doing here and why?" he asked.

I told him we were looking for the Death Worm. At one point two very glamorous looking young women burst in to the room. The major looked embarrassed and quickly shooed them away. Throughout the interrogation, I remained polite and helpful, although I felt seriously annoyed. You see, we had met this same colonel only a few days before near the Gobi National Park. At that time, he had been in civilian dress; he knew that we were heading this way. I found the whole episode bizarre.

Meanwhile Bilgee had been in touch with Byambaa, who was pulling some strings in Ulan Bator. Apparently the colonel wanted us to pay a fine for being in the area. Byambaa, I would later learn, had phoned his boss in Ulan Bator, a general who had told the colonel to release us and quickly!

And that is just what the colonel did.

"He wants you to sign this," said Bilgee.

"What is it?" I said, and asked Bilgee to read it back to me. It was basically a short diatribe saying that I was sorry I'd caused him an inconvenience and I now recognized I'd been in the wrong place at the wrong time.

I was happy with this, as it meant we could be released and I wasn't writing a confession to any crime; otherwise, I wouldn't have signed it.

"Good luck," he said as I left his office. I smiled as I left his office and let out the most virulent fart I've ever produced. I wanted the colonel's last memory of me to be a vodka and mutton special.

We rejoined Andy who was waiting for us at the jeep. From then on we had to report our movements, mainly by phoning the base from post offices along the way. Apart from this inconvenience, we were in high spirits and excited about the witness statements we had gathered. I used a tabulated chart to cross-reference features from those accounts with their locations. One of the things that struck me was the high degree of correlation between reports of the creature's physical description. All had seen a maggot-like creature, which mostly referred to its burrowing action. All described it as being snake-like. One particularly graphic account came from a man who had mistakenly touched a Khorkhoi; his arm had started to burn, so he had stuck it in a bag of cooling airag, which turned green from the poison.

Armed with this information, we decided to try our luck at three locations close to our current position given the time we had available. We resolved to trek at different times of the day in an effort to find evidence of the creature: once first thing in the morning for two hours, then after breakfast for another two hours, again in the afternoon after lunch, and one more time in the early evening for a minimum of six to seven hours total. It turned out to be tough going. At altitude in the desert, you begin to feel that you are walking in a suit of armor. As you become increasingly hot and begin to suffocate, your mental processes start to slow and your words become slurred.

Our tent didn't help much either. Before leaving, we had discussed buying a specialist tent designed to handle high altitudes, UV exposure, and heat, but as it cost at least 400 pounds, we simply couldn't afford it. As the Mongolians offered to provide us with a tent, we decided to settle for that. Unfortunately for us, it appeared to be a child's tent purchased from the Mongolian equivalent of Woolworth's and was decorated with pictures of clowns and teddy bears. One clown in particular stood out and grinned at me at night as I lay in my bed: the sinister looking Charlie. To make matters worse, sometimes we had to spend hours inside our tent, simply to get away from the baking heat outside, which at times as high as 49 degrees Celsius or 120 degrees Fahrenheit. As a result, I now have a pathological hatred of clowns.

Nevertheless, the tent stood up to some pretty brutal treatment. It survived a major sandstorm and looked in better shape than we did by the end of the expedition. The clowns, unlike us, were able to sustain their maniacal grins even on the toughest of days.

At the first camp, we found plenty of wolf tracks, including some that appeared right up to one of our tents and then turned away. We also saw lots of airborne camel ticks, but no joy on the Khorkhoi front. Before we moved on to the second camp, we spent some time at a local village school, giving out postcards of the U.K. I also had the second worst meal of my life while in the village of Camel Borsch. I'd had camel vodka on the way down to the Gobi and found it surprisingly good, but the meat, mixed as it was in a greasy pancake with lumps of hair and gristle, did nearly make me puke. It wa served swimming in its own "sauce," a sort of bright green fatty substance, the origin of which I never did discover. That's probably a good thing.

Upon reaching the second camp, I immediately felt this was the perfect habitat for the creature, and indeed it has been sighted there many times. Unfortunately, it was also home to thousands of biting flies. One day I acquired more than fifty bites; even toxic combinations of various proven insect repellents provided no relief.

Nevertheless, it was here that we celebrated Andy's birthday with the by now customary evening vodka. At night Jagaa was always the first to go to bed, sleeping in his jeep, followed in turn by Andy, and then Bilgee and I into our respective tents. One thing had puzzled me though, and I quizzed Bilgee on it that night.

"Bilgee, why does Jagaa's jeep shake at night?" I asked.

"Oh, he is probably thinking of the postmistress," replied Bilgee nonchalantly. You can imagine our reaction to this – we both fell about laughing.

"What's the matter with you, Adam," said Bilgee defensively. "He is a man, after all."

"Fair enough," I spluttered between roars of mirth.

I did, however, check that back seat carefully before I sat on it the next time.

On our last afternoon there, I had instructed Jagaa to break camp while Andy, Bilgee, and I went on our final trek. As we returned to camp, we found Jagee waving his arms excitedly. It turned out that one of the dreaded spiders had taken refuge under our tent, and when he had moved it, the creature had charged him. It was now lodged under our water container and as Bilgee moved it, the spider behaved exactly as the old man had described: it seemed to raise it legs and charge straight for me! I quickly dodged out of its way.

"Kill it," cried Bilgee.

"No," I cried. "Give it a chance."

It then turned on Bilgee and managed to get up his leg. One bite from the spider would be highly poisonous and, considering the remoteness of our location, very dangerous. Bilgee managed to shake the beast off without getting bitten. Undeterred, the creature then turned on Andy.

"Kill it!" I cried.

Shortly afterwards, we broke for our final camp. On the way there, we stopped at the Ger of an old man who had been part of a geological investigation in 1948. He was an enormously enigmatic character and recounted his tale over tea, hunched down in his Ger, punctuating the story with the occasional snort of snuff.

He revealed how one of the geologists had first spotted a Khorkhoi and had asked him exactly what the creature was. Both of them had been nervous and left the creature alone. The following day, the survey team saw two more – and promptly burned them.

When I asked the old man why they had done so, he turned towards me, stared me straight in the eye, and said, "Because we were afraid. I am afraid of this place. It's a bad place. Since then I have only been back there twice in my life. You will see."

The Mongolians are superstitious people, so Jagaa and Bilgee did not look forward to our arrival at the last camp with anything but trepidation. My main concern was having enough water to stay there. Gobi well water is very salty and we were some distance away from the nearest well. We were on a large ridge of jutting sandstone, and as we reached the bottom of it, we could feel ourselves cooking in the heat, as if we'd fallen into a giant frying pan. At the bottom were dozens of animal carcasses, casualties of the previous year's droughts.

I have to say the place did feel very unnerving. After spending a few days there, I could sympathize with the old man's reservations about it.

We were now running out of time and water, and reluctantly we had to turn back without finding the Death Worm.

The journey towards the steppes was through desolate arid land, which looked in places very much like the surface of Mars. On one occasion we had to stop, as our route took us past some very shady looking illegal gold miners. We were concerned about being robbed, and as our jeep pulled up to them, I reached for my trusty knife.

Bilgee and I got out of the jeep and prepared to approach them.

"What are you going to do if they attack you?" asked Andy, who was fiddling with his camera case.

"Stab them," I said.

"Oh, fair enough," said Andy nonchalantly and carried on fiddling with his camera case.

The miners proved to be no danger, however, and gave us some directions. Just before we entered the nearest village we encountered some Mongolian police who had parked their jeep by a well. They'd captured some horse thieves and were making them turn the well to get everyone, themselves included, a drink.

Arriving in the village, we went to visit the house of a Naddan champion wrestler, who was a friend of Bilgee's. In one of the stupidest decisions of my life, I volunteered to wrestle him. Given my complete inexperience, this was akin to someone who had never boxed volunteering to take on Muhammad Ali. Even my tough Mongolian team, who encouragingly referred to me as "big boss," looked a little worried.

Of course, he beat me. I didn't have any idea of the techniques required, nor did I have the years of training, but they admired my guts and laughed as I repeatedly hit the dirt. The wrestler's brother offered to kill a goat in celebration, telling me that I would be a wrestler if I lived there.

A few days later, we arrived back in Ulan Bator. Byambaa was there to greet us at our hotel. As we emerged dirty and tired from our jeep, he said: "Tonight I have a special treat for you Adam."

"What is it?" I enquired.

"Lets just say this," he said," I won't be bringing Ashnaa." Ashnaa was his girlfriend.

A few hours later, having showered, changed, and eaten dinner, we were off to the bar. At first it seemed like any other bar. There was a drinks area and a dance floor, but no one appeared to be dancing.

"Come on," said Byambaa, urging me on to the dance floor. The music started pumping out a loud techno beat. Now, I was feeling awkward at this point, but not as awkward as I was about to become, for from out of the wings of either side of the dance floor emerged about a dozen young women clad only in thongs, stockings, and suspenders.

"This is incredibly fucking seedy," I said to Andy, even though, I have to admit, I was bemused by the whole thing.

"Yeah," he said, "but we are here, so we might as well stay!"

And stay we did. I ended up drinking with Byambaa and Bilgee, while Andy ended up being hustled at pool by one of the girls.

At the end of the expedition, arriving back in Ulan Bator, Andy and I were given a marvelous present – a four banded, silver, puzzle ring, each band representing one of our team. They were handmade by Jagaa's brother and represented the lifelong friendship we had forged together. It would serve as a moving reminder of a fantastic experience, one in which we had traveled to areas never before seen by Westerners, even if we hadn't found the Death Worm.

And what of this mysterious creature we had gone so far to find?

Well, I found no scientific proof of its existence. I did, however, find a number of highly corroborative stories from eyewitnesses. The habitat and remoteness of the area are also consistent with a creature of this kind, the ecosystem could support it, and the vast and remote areas where it is said to live might explain why it has remained unknown to science.

At this point, I can only echo the sentiments of Roy Chapman Andrews: "If the faith in its existence was not so strong and widespread… and if everyone did not describe the animal exactly the same way, I would believe it to be an idle myth."

Chapter Eleven

Russian Bigfoot

I found myself back in Mongolia just a few years later. This time I wasn't hunting the Death Worm in the barren southern wastelands of the Gobi. I was headed to extreme Western Mongolia, towards the border with Kazakhstan, where stories of the Almas, the Mongolian version of the Yeti, had abounded for centuries.

The following account of the Almas comes from Hans Schiltberger, a prisoner of the Turks, and dates back to the 15th century. "Tschekra joined Egidi on his expedition to Siberia, which it took them two months to reach. In the country there is a range of mountains called Abrs which is thirty-two days journey long… In the mountains themselves live wild people, who have nothing in common with other human beings A pelt covers the entire body of these creatures. Only the hands and face are free of hair. They run around in the hills like animals and eat foliage and grass and whatever else they can find. The Lord of the territory made Egidi a present of a couple of forest people, a man and a woman. They had been caught in the wilderness, together with three untamed horses the size of asses and all sorts of other animals which are not found in German lands therefore I cannot put a name to."

This account impressed me for two reasons. Firstly, Schiltberger reports what he saw with his own eyes. Secondly, he refers to Przewalski horses, which were only rediscovered by Nicholas Przewalski in 1881. Przewalski himself saw wild horses in Mongolia in 1871.

Then we have an 18th century Tibetan Book of Medicine, which simply lists the Alams as one of the indigenous animals that abound in the area – no mythical animals are mentioned in the report.

Unlike its more famous Himalayan cousin though, very little Western field research seems to have been conducted on the Almas.

One exception is the work of British anthropologist Myra Shackley, who gave the following descriptions of the Almas in her book, *Still Living? Yeti, Sasquatch and the Neanderthal Enigma*: "Their jaws are massive, their chins recede, their eyebrow ridges are very prominent compared with those of the Mongals and the females have long breasts which, when they are sitting on the ground, they can toss over their shoulders to feed the infant that is clinging to their back... Some reports also say that the feet are slightly bent inwards and other people note that the Almas can run very fast and are incapable of making fire. They are frequently reported to be primarily nocturnal, timid, unaggressive and unsociable. They eat both vegetable and animal food, primarily small animals. They do not have an articulate language and they seem incapable of uttering more than a few word-like sounds."

Shackley wondered whether the Almas could, in fact, be a remnant of Neanderthal man. Though the idea of a pocket of Neanderthal men holding out in a remote area, a last surviving enclave, struck me as an unlikely notion, I did consider the possibility that some type of hominid species could still be living there. After my experiences in Sumatra, I had no doubt that some unique undiscovered bipeds did still exist, and so we began some investigations with the help of my earlier Mongolian contact, Byambaa.

In the course of preparations, Cara Biega of *National Geographic* got in touch with me and expressed an interest in coming along. I liked Cara, so I readily agreed. This time Andy wouldn't be coming with me, as a career in the Navy beckoned. I much prefer traveling with a friend and sharing experiences to introspective self-discovery.

Byambaa and Bilgee had been hard at work preparing the ground for when I arrived. They had established that there were living Przewalskis in the far west, near the town of Hovd. They had also found some living Mongolian scientists who had researched it. I was delighted to have some decent leads to follow up. Since I had arrived a few days before *National Geographic*, I took the time to ensure all the logistics were in place and, most importantly, interview a key expert.

I met Professor Navaan at an excavation he was conducting at the Touil Khan temple in the countryside just outside Ulan Bator. I taped my interview with him, using my little video camera, as the TV crew had not arrived yet. Professor Navaan was an enormously energetic man despite being in his 80s. My encounter with him was to prove quite startling. A professor of anthropology, he had, among other things, spent some time researching the Almas.

"Is it real," I asked him.

"Yes, undoubtedly it used to exist," he said, nodding sagely.

"What do you think it is? Some scientists have postulated it's a Neanderthal man," I said.

"No, it's not a Neanderthal man. Nor is it an ape – it definitely walks erect, but it has none of the trappings of civilization you might expect from a Neanderthal man."

"Well then, what do you think it is?" I said.

"It's not an ape; it's not a Neanderthal man; its something in between," he replied.

"So a new and distinct species," I said.)

"Exactly," he replied, as he sipped on his beer.

Possibly the most telling revelation came at the end of my interview with him.

"Does it exist now?" I asked him.

"No," he said, "the last ones died out in 1921. I went looking for them in the sixties and seventies. I used dogs, trackers, everything; there were no sightings. We found nothing. I say 1921, as that was the last credible sighting by a Russian colonel."

I was surprised that a professor of anthropology would be so assertive in his conviction that the Almas had existed as recently as 1921, especially on camera.

As I prepared to leave, he said, "Good luck. I hope you find it. And if you do, you'll be the most famous man in the world. But I think you'd hate that; you love your freedom too much." He laughed as I left.

That night I tucked in to a juicy steak at a restaurant with Bilgee and Byambaa and caught up on the news. It turned out that the man who had had us arrested in the Gobi on my previous trip had been sacked. Apparently, he had been running some sort of con, trying to exhort money from Chinese businessmen. When the Mongolians realized what he was up to, they promptly fired him. I wasn't sorry; he was a bastard.

The following day we picked up Cara, the film's producer, and Rob, who was the film's director, at the airport. They had been filming in Moscow for the program, interviewing eyewitnesses and Russian Almas researchers. While there, Rob was bitten by a vicious stray dog and unfortunately needed treatment with large doses of vaccine in case the beast had contracted rabies. All things considered, Rob was in remarkably good humor, and we all went to the beer tent in Ulan Bator that had been especially set up as the World Cup was on.

We went through our plans and discussed the research we had gathered. One thing we all agreed on was that most of the sightings were in an area just a few hours drive from Hovd. We were to travel there and meet the witness of one of the most recent sightings, a man called Ulzi. But first we would met Jagaa in Hovd; he was running a team of drivers who had already been dispatched to travel across the country and rendezvous with us at Hovd. We flew there to make best use of the time. I was pleased to hear that *National Geographic* had hired the services of a cook for as long as they were out there; Bilgee is a really great bloke but a really lousy cook. Cara and Rob would spend ten days or so on location and then return to the U.S. I would then remain with Bilgee and Jagaa and continue the hunt, eventually returning via land.

Arriving at Hovd, Cara asked me about the wildlife. "Is there anything dangerous I should know about out there?" she said.

"Not really," I said, "although there does seem to be a lot of wolves in the wilderness."

"Wolves?" said Cara. "You never told me anything about wolves, Adam." Cara was clearly alarmed.

"I'm sorry," I said in an apologetic tone, but I was unable to hide a smile. "But they are really unlikely to come anywhere near you."

That night I sat in my Ger, which I shared with Bilgee. We discussed what we were going to do after the *National Geographic* team left. As ever, we had to balance our desires against time and money, but eventually we agreed on a feasible plan. I wanted to go southwest first to the Kazath area and explore their stories and then go south. I then hoped to establish the possible range of the creature, should it still exist. I would have to do some hard bargaining with Jagaa, especially as we would be traveling to another border area. Then suddenly I heard a woman's scream coming from the direction of the tower block.

Alarmed, I ran towards it and found Cara standing in the middle of the path. "A wolf, a wolf," she was shouting. A small white dog then came out of the gloom and approached us, wagging its tail.

"Wolves don't wag their tails on approach, Cara," I remarked as blandly as I could.

The following day we traveled to see Ulzi. I had raved to Cara about how great a driver Jagaa was, but as he backed into the compound where Ulzi lived, Jagaa managed to clip a donkey at about five miles an hour.

"Great driver," said Cara. The donkey walked away undisturbed by the experience.

I enjoyed meeting Ulzi, who was an extroverted and chatty man. He was in his early sixties and full of vigor. He was a typical Mongolian nomad with a wife and family. He recounted how, in the 1990s, he had been driving some Mongolians to a scenic spot for a picnic when he had seen an Almas by the river, presumably drinking. What struck him most was that it stood upright. When he first saw it, he though, *just like a man*, but when it ran away, it scurried off *just like an animal*. I was intrigued by this and was keen to visit where he had seen the Almas.

When we got there, I saw a large verdant pasture by the river and up above were some small caves. I interviewed Ulzi there, a short interview for the cameras and a longer detailed interview for my own research. I had been working as a British civil servant in the courts and had cross-examined people every day over the last few years. There were two key points that particularly struck me.

First, I asked him if he could have been mistaken.

"Not a chance," he said, firmly. "I am a hunter. I have hunted here for years; I know all the animals. I would not make such a stupid mistake."

The second point that struck me was the creature's movement. He had first described the Almas as looking like a man, as it stood on two legs with an arched back. But when I got him to describe how it ran away, he clearly stated that it was like an animal. The scurrying motions he described reminded me more of a chimpanzee in flight. What was abundantly clear, though, was that he was not describing a Neanderthal, more a unique type of ape, just as Professor Navaan had postulated.

Ulzi indicated that he had trekked further into the wilderness from that point many times. But beyond that point, vehicles could not go because of the terrain. I was definitely interested in going back there again. But our short schedule meant that we had to plough on.

We now traveled higher up the mountain, where there was still snow in some places despite the fact that it was now June. We settled on camping in a high pasture area, where there were a number of families in their respective Gers. As we talked that night, I heard a crackle of gunfire ring out across the valley. I looked at Bilgee. "They are shooting the wolves," he said. "There are many wolves in the mountains here."

The following day we did a tour of the area, interviewing many of the local people. It soon became very clear to me that while the old people were familiar with the stories of the Almas, most of the young people knew little of the legends. One particularly engaging old man named Terbat told me how people used to claim they had seen it when he was a child, but he had not heard of anyone saying that they'd seen one for years. He also told us some of the legends surrounding the Almas. One story in particular piqued Rob's interest. Apparently, a nomad had been tending to his goats one day when a female Almas had abducted him. She kept him trapped in a cave and fed him raw meat. She forced him to have sex with her and they produced a child. One day, the man managed to escape because he was able to cross a large river and the Almas, being afraid of the water, could not follow him.

"I wouldn't mind being abducted," joked Rob.

"Yeah, but I like my women without hairy backs," I replied.

Joking aside, there were some very obvious recurring insights on behavior here. Even if the stories were based on myth, the descriptions and qualities attributed to the Almas were similar. Everyone described a creature that was hairy all over, apart from its face, which was described as being "human-like." None of the stories gave the Almas even the most primitive technology, e.g. the manufacture of spears or the ability to make fire. None of the descriptions sounded anything like a Neanderthal to me.

Much of the rest of the filming was given to a "stakeout scene" around a large cave to the south. That night Bilgee again recounted his aversion to vegetarians. "They are nothing but grass-eaters," he said.

"Wouldn't you just like a cheeky carrot?" I said, goading him.

"We are Mongolians. We are men. We eat meat," he replied defiantly.

I reflected that, according to Bilgee's theory, the hoards of the Great Khan could have been stopped by crowds of angry greengrocers flinging broccoli or carrot sticks in their general direction. Yes, it was time to go to bed.

The next day we turned back towards Hovd to ensure that Cara and Rob made their flight back to the U.S. Camping that night we discussed the merits of dung on the fire.

"How do you decide which piece of dung is the best to put on the fire?" I said to Bilgee.

"Simple," he replied. "If it's grey, it's okay; if its brown, leave it down."

We all had a go at chucking some on, including Cara, who fashioned some giant chopsticks out of two branches, to chuck hers in, which made the Mongolians laugh.

Back in Hovd, I interviewed a Tibetan monk and the head of the monastery, to see if he had any knowledge of the Almas.

"I'm not sure if I believe it," he said to me, "but I know Ulzi, and he's no liar, so it must be."

Cara was always very well organized before a trip and her researcher had come up with an account from a local paper. Apparently, the story involved a Kazalch man who had found some strange footprints. In his account, Ulzi had described finding footprints on the muddy riverbank after the creature had run away. Like Ulzi's description, the key unusual feature of the footprints found by the Kazalch man was that they appeared to have "long toes."

We waved Cara and Rob off at the airport, and Bilgee, Jagaa, and I headed off into the wilderness. The first camp we found was unfortunately full of mosquitoes and we had a rough night with them.

The following morning I saw a drowsy Bilgee emerge from his tent.

"Bastard," he said, as he swatted a mosquito that had landed on his neck.

Jagaa meanwhile had become less vain now that the camera crew had gone. Both Bilgee and Jagaa were hard men; both had several teeth missing from fighting. What amazed me, though, was that Jagaa preening himself in his jeep mirror before he did any filming scenes. At one stage Bilgee even said, "I think I need makeup." But I'm sure he was just joking.

We now headed further into the wilderness proper – past the mysterious "Deer Stones," the ancient stone monuments that hold as much awe and mystery for the Mongolians as Stonehenge does for us in the U.K.

We had by now arrived near the border, which means we had to stop and ask for permission to carry on. Jagaa and I waited in the jeep, a little uncomfortably, while Bilgee went in and spoke to the major who ran the base. This time, though, there was no problem.

After traversing a beautiful but difficult steep-sided valley for a few days, and crossed a river, we eventually arrived at the town of Bulgan. I had wanted to visit this place because it was marked on the map as being the "Mountains of the Almas." Though I hoped to gather new information and leads here, I was disappointed. The "mountain" did indeed exist but proved to be little more than a hill on the edge of town. No one had seen the Almas. "They are little more than stories to scare children," said one man gruffly when we asked.

Moving on we arrived at the town of Altay, which was deserted. There were no voices, no children playing in the park, nothing. We went to the post office and found it open. There we learned that almost the entire town had moved out for the summer due to the large number of biting insects that had moved in to attack them!

So we headed north towards Kazah. We were forced to do a perilous river crossing. We had crossed several rivers on our way down the valley, but this one was particularly serious and we weren't sure whether we'd manage to make it across. Freezing meltwater from the mountains had turned what was normally a gentle river into a vicious torrent. Jagaa reasoned with me that if we didn't cross, we would lose half a day or more trying to find a new route, and that given our time constraints, we should give it a try. We could always give up and turn around if it didn't work out.

So we ploughed on into it. We were fine as long as the bed of the river was constant. Then we hit a dip and the hood of the jeep went straight in to the meltwater. The engine cut out. I looked down and could now see the meltwater coming up above my boots.

"I think we're going to have to get out of here," I said to Bilgee.

"Yes," said Bilgee.

I prepared to ditch my kit and make a swim for it. My mind whirring, I reasoned I could swim with the current and then pick my moment to break across it and swim for the shore. I would have to be quick, though, for the river was little more than ice water and I would freeze. But at that point the jeep sprang to life and a highly stressed Jagaa, who had been furiously pumping the ignition, breathed a sigh of relief as he guided us safely over to the other side. It's a good thing the Russians build their jeeps tough, otherwise we wouldn't have made it. After we managed to recover from the experience, I asked Jagaa if he was a good swimmer had worst come to the worst.

"I swim like a rock, Adam," he said. Jagaa came from the Gobi and had never learned to swim.

Eventually, we arrived at the Kazah town of Bulgar. The Kazah Mongolians are a distinct minority ethnic group with their own customs and traditions. They are predominantly Muslim and are perhaps best known for their eagle hunters. Around October, the hunters take specially trained birds out, staking large game. The eagles are superb hunters, and those families that keep them take enormous pride in their birds. The eagles are also famous for kidnapping their wives!

The nomads among them also have distinctive Gers, which are higher with taller roofs and chimneys than those of most Mongolians.

We arrived in town on the day of the local Nadaan Festival. I found the Kazah's to be friendly. Most were aware of the Almas and did not treat the accounts of it with scorn. "Oh, it's a real animal," came the consistent reply. Several people also talked of how the legends said the Almas made a "high whistling sound" that was quite distinctive and unmistakable.

As we headed further north, the number of encounters with, and stories about, the Almas, seemed to increase. Arriving in Monyilian we made contact with the son of a man who had seen Almas footprints. A keen amateur mountaineer, he had seen a distinctive and impressive trail of humanoid footprints while climbing one day. There was no one else around, and he had no doubt about what he saw. However, he had been afraid to talk about what he had seen.

"Why?" I asked his son.

"He was a teacher and it was Socialist times," came the reply. It did make me wonder if others had seen the creature and, through fear, had never dared to talk about their sighting. For example, Stalin had exercised vicious suppression of shamanism, which was popular in the north of Mongolia, and accounts of the Almas might very well have been confused with mysticism, given the legends surrounding these creatures.

Once again we were running out of time. We had little choice now but to begin the long journey back to Ulan Bator.

We first headed to Lake Dogon, which is huge. Arriving there, we met up with a local family Jagaa knew. The husband and wife cracked open a bottle of vodka for us after we had a refreshing dip in the lake, and it soon became apparent that I'd made a new friend. Both the man and the woman were wrestlers. The woman, however, seemed quite keen to show me the art.

"I will take you in to my Ger and you can try on a costume and I will show you," she said.

"She likes you," said Bilgee, laughing. "Maybe you should go."

"I don't think this romance has time to blossom," I replied.

The following morning, after much teasing from Jagaa about the woman wrestler, we continued our journey. Our plan was to travel west and eventually pick up the tourist route and go to Karakorum. On the way we stopped off at the "Cave of the Almas." On the inside of this cave in the countryside were hand-carved prehistoric drawings of an Almas. Bilgee and I climbed up the promontory to the cave, while Jagaa remained behind with the jeep. While in this cave, we came across sacks of wool. "Somebody smuggled them in here," said Bilgee. "It looks like cashmere, so it will be pretty valuable."

After a short diversion to visit the monastery of Karakorum, which was quite impressive, we eventually arrived in Ulan Bator. Byambaa told me the story of someone who had come back from a recent journey into the countryside having found some stolen wool, thinking it was expensive cashmere. In fact, it turned out to be very cheap camel wool, and was barely worth a few bucks. That person had had to endure some severe teasing.

That night Bilgee and Jagaa presented me with a lovely silver bowl. "This is for drinking your vodka out of. Please think of us when you do."

"Thank you both very much," I said. "I will." And I do.

On the plane home, I took some time to reflect about the Almas. Does it exist? I believe it did. Legends aside, credible eyewitness accounts date back centuries from all sorts of people. Military men, such as the Russian Colonel who saw it in 1921, would have been very careful not to exaggerate their reports.

What is the Alams? I am certain that is not a Neanderthal. Even if an isolated group of Neanderthals had undergone a period of cultural recession, there was no evidence to suggest that the Almas had anything other than a level of technology equitable to that of chimpanzees. None of the legends, for example, describe them as even having the ability to make fire.

Then there is the physical description of hair all over its body but with a humanoid face. I agree with Professor Navaan, the Almas were/are a unique species, a homo erectus no doubt. People often find it hard to grasp the fact that different species of man lived together for many thousands of years. Only relatively recently has our species come to dominate. The Almas seem to be an isolated remnant of that earlier time.

Does it still exist? I find this question much harder to answer. Recent encounters are few and far between. There are a few sightings of footprints here and there, but only one physical sighting that I know of in the last ten years (the one by Ulzi). Even though the area is very remote and inaccessible, the number of sightings of Almas is worryingly low. Contrast that, for example, with the high number of sightings of the jungle-dwelling *orang-pendek*.

So it seems to me that the Almas are a wonderfully unique species, but a species that is probably past the point of extinction. I traveled over three thousand kilometers in Mongolia in my quest for the Almas, and it's quite clear that the only viable range remaining must be in a comparatively tiny pocket of the Western mountain range – no more than one thousand kilometers square or so. If there are a few Almas left, they will be very hard to find, given the terrain; I'd say it's almost impossible, in fact. It would be a dangerous journey. If anything went wrong, given the remoteness of the area, I'd probably be doomed. But if there's a chance...

Maybe one day I'll give it another go!

Epilogue

Ever since I was young, I hoped that Bigfoot would jump out of the darkest depths of the forest I was trekking through, or that the Loch Ness Monster would rise out of the water as I sailed a little boat across its waters.

Even as an adult I always started out with the belief that these creatures were real. But in my travels, whenever I haven't found anything, I've said so. I really wanted the Loch Ness Monster to be a giant pleisosaur or something, and for there to be a viable population of these creatures living in the Loch today. But I'm of the opinion that there just aren't any such creatures. Personally, that's disappointing, but I have to tell the truth as I see it.

I did see the Seljord Serpent, Norway's Nessie, and that really surprised me. It was a fascinating and thrilling experience that will stay with me for the rest of my life. I've also gathered convincing evidence for the existence of the *orang-pendek*, the Sumatran Yeti. The analysis of hairs and footprints that I found and brought back has led some credible scientists to agree that there is indeed an unknown primate living in the jungles of Sumatra.

Of all the cryptids I've looked for, it is the situation facing the *orang-pendek* that probably upsets me the most. I know it is out there; I've been out looking for it four times. The reason it upsets me is because I know that with the increase in the local population and environmental pressures, time for this creature is running out. I've done my best to help, but I'm desperately worried that the *orang-pendek* will die out in my lifetime. I just hope that the Indonesians come to recognize what a national treasure the *orang-pendek* really is and help it to survive.

In the meantime, I'm going to carry on, seeking out unknown animals around the world. I won't stop until I die.

Expedition Gallery

The author driving a speedboat to the location on Lake Seljord, Norway, where he and Andrew Sanderson saw "Selma." (Photo by Andrew Sanderson)

Lake Tele in the Congo is the reputed home of the legendary mokele-mbembe. (Photo by Adam Davies)

The villagers of Boha in the Congo celebrate the safe return of our mokele-mbembe expedition team. (Photo by Adam Davies)

Adam Davies, Andrew Sanderson, and the "Tiger Team" at the end of their successful 2001 orang-pendek expedition. (Photo by a Tiger Team member)

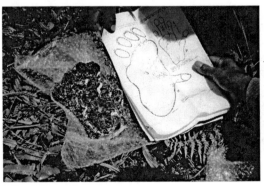

Sahar, our forest guide, is shown comparing a drawing of an orang-pendek footprint to an actual footprint we found in 2001, moments after it had been extracted from the ground. (Photo by Andrew Sanderson)

The tiny object in the middle of the image is the Jeep we used during our expedition in search of the Almas. The vastness and remoteness of the Gobi Desert is apparent. (Photo by Adam Davies)

Making camp in Mongolia while searching for the Almas. This is where a local nomad named Ulzi claimed to have seen one of the creatures. (Photo by Adam Davies)

Hair Scans

 Chimpanzee

 Ebony Leaf Langur

 Gorilla

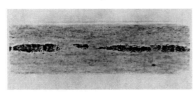 Human

 Orangutan

 Pig-tailed Macaque

 Red Leaf Monkey

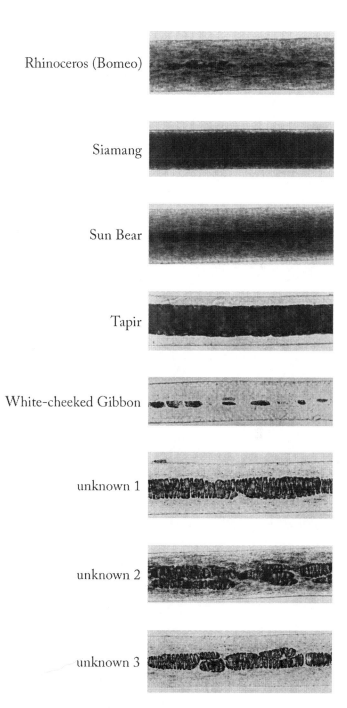

Acknowledgments

There are a number of people I would really like to thank.

For making this book possible: Loren Coleman and Patrick Huyghe.

For their help in organizing expeditions: Jacques Mayambika for The Congo; Byambaa, Bilgee and Ashmaa for Mongolia; Debbie Martyr and Jeremy Holden for Sumatra; and Jan Sunberg for Norway and Loch Ness.

For my friends: Andrew Sanderson, for his logistical help; John Macdonald, for the construction of the initial website; Keith Townley, for making me laugh.

And also for the guys at the Centre for Fortean Zoology: Jon Downes and Richard Freeman, for helping to carry the torch so well in the UK. And for John Everatt, for his help in editing this book.

Adam Davies
Manchester, U.K.
March 2008

Chronology of Expeditions

January/February 1998
First trip to Sumatra – Orang-Pendek

July 1998
First trip to the Congo– Mokele-Mbembe

August 1999
Norway – Selma

October 2000
Scotland – Loch Ness Monster

November/December 2000
Congo 2 – Mokele-Mbembe

September 2002
Sumatra 2 – Orang–Pendek

June 2003
Mongolia – Korkhoi

September 2004
Sumatra 3 – Orang-Pendek

June 2006
Mongolia – Almas

September 2007
Sumatra 4 – Orang-Pendek

Printed in the United States
109199LV00001B/10/P